AN INTRODUCTION TO PULSE N M R SPECTROSCOPY

PULSE NUCLEAR MAGNETIC RESONANCE

SPECTROSCOPY

An Introduction to the Theory
and
Applications

THOMAS C. FARRAR
Department of Chemistry
University of Wisconsin
Madison, Wisconsin

The Farragut Press Chicago

THE FARRAGUT PRESS
P.O. Box 5102
Madison, Wisconsin 53705

Library of Congress Catalog Card Number : xxxxxxxxx

Printed in the United States of America

ISBN 0-917903-00-5

CONTENTS

PREFACE

The first commercial pulse nuclear magnetic resonance (NMR) spectrometers became available a little over fifteen years ago. By today's standards these early instruments were primitive and very expensive devices.

In the past fifteen years pulse NMR spectrometers have improved greatly. This improvement has been the result of enormous changes and technological advances in both radio frequency (r.f.) electronics and, of course, in the almost unbelievable advances in digital electronics and computers. Ten years of software development have also brought about great improvements and have made the spectrometers much easier to use. Experiments which ten years ago could be executed only by Ph.D.-level NMR spectroscopists can now be carried out by any schoolboy. These improvements have, in turn, had an enormous impact on chemical research. For example, it is now routine to obtain very high quality proton NMR spectra in a minute or two with only a few hundred micrograms of sample. Or one can obtain carbon-13 NMR spectra of a few milligrams of sample in just a few minutes. In addition, one can now obtain carbon-13 spectra of only those carbon atoms with three attached protons, or of only those carbons with two attached protons, and so on. And the rate of exciting new developments continues unabated. Just in the recent past a whole new class of 2D-FT-NMR experiments has been developed which has afforded a 100-fold improvement in sensitivity and now provides a wealth of important new information on the structure of complex molecules.

The purpose of this book is simple: to provide a readable introduction to pulse NMR spectroscopy and to Fourier transform NMR. The approach here has been to describe as simply and clearly as possible the different basic pulse experiments and to give simple experimental examples of their applications. I have tried to provide the necessary theoretical background in a simple and accurate manner. In some cases simple classical pictures have been used, rather than esoteric or difficult and rigorous quantum mechanical derivations.

The selection of references is difficult for a book of this type. I have chosen those references which were, in my view, either seminal papers in an area or were early, readable papers with a clear presentation of the basic principles of an experiment. The reference list in no way attempts to review the literature. My apologies to the authors of the many excellent papers and books which are not included in this brief list.

The first two chapters review the fundamentals of NMR spectroscopy and introduce the basic pulse methods. Chapter 3 summarizes the salient features of pulse spectrometers, while Chapter 4 explains, in words as well as equations, why and how nuclei relax. Chapter 5 is devoted to: (a) the details of Fourier transform methods, (b) important concepts and uses of the computer, (c) computer control of the spectrometer, and (d) acquisition and manipulation of the data. Chapter 6 describes in some detail the theory and application of some of the more recent developments in multiple-pulse, multiple-phase experiments, such as the sensitivity enhancement-spectral simplification techniques, that is, the INEPT and ERNST experiments. Chapter 7 gives an overview of two-dimensional Fourier transform experiments with a few simple examples.

Modern NMR spectroscopy is an absolutely indispensable tool for the practicing chemist. In order to make the optimum use of this increasingly powerful tool it is essential that some of the more important fundamental concepts be mastered. I hope that this introduction to current NMR theory and applications will provide the reader, whether he is a practicing chemist or a beginning graduate student, with a solid working foundation. And I hope that this small volume will help stimulate further application of pulse NMR techniques to a wide range of chemical problems. This simple introduction should, in addition, provide the reader with a foundation for further study of the current literature and more advanced texts. The elegant and lucid works by Slichter and by Abragam still remain monumental works which provide detailed treatments of NMR theory at a very sophisticated level.

Thomas C. Farrar
Department of Chemistry
University of Wisconsin
Madison, Wisconsin
February, 1987

ACKNOWLEDGMENTS

It is not possible to properly thank all of the very many friends, students, and colleagues who have helped me in many ways in the writing of this book. Special thanks are, however, due to a number of those who gave generously of their time to help make this book possible. Bruce Adams, Jim Cooper, Cecil Dybowski, Joe Noggle, and Don Woessner read the entire manuscript. They caught numerous errors and made a number of suggestions which, I think, have made the book clearer and easier to read. Bruce Adams also provided a number of the NMR spectra that are given as examples throughout the book, and he spent many hours discussing how to present clearly to beginning students some of the more difficult concepts. I am especially indebted to the editor of the Farragut Press, Ms. Roberta Gutman. Although she is a very tough editor, she has been very patient and it has been a joy to work with her. I thank Christian Farrar for help with the figures and Michael Farrar for help with typing the manuscript.

SYMBOLS AND ABBREVIATIONS

SYMBOLS

A	spin-coupling tensor
A_{bx}	spin-coupling constant between nuclei b and x expressed in units of radians per second
A_∞, A_τ	amplitudes of the Free Induction Decay signal in a spin-echo experiment
$\mathbf{B}_o$	direct current magnetic field
ΔB_o	inhomogeneity of B_o
$\mathbf{B}_1$	macroscopic r.f. field
$\mathbf{B}_{eff}$	an effective magnetic field
B_t	total magnetic field at a nucleus
B_{loc}	local magnetic field at a nucleus
B_d	local field arising from static magnetic dipoles
C	capacitance
C_{eff}	effective spin-rotation constant for molecules in the liquid state
$C_\parallel$, $C_\perp$	components of **C**, the spin-rotation tensor, parallel and perpendicular to the symmetry axis
C_{3v}	threefold symmetry axis
c	velocity of light
C	spin-rotation tensor
D	separation between field gradient pulses
d	duration of field gradient pulses
D_c	diffusion coefficient
$D_{c\parallel}$, $D_{c\perp}$	components of D_c parallel and perpendicular to the symmetry axis
ΔE	separation between two energy levels

E_a	activation energy
E_c	interaction energy
e	charge of the electron, also a constant (≃ 2.71828...)
e^2qQ/h	quadrupole coupling constant
$F(\omega)$	Fourier transform of f(t)
$f(t)$	Fourier transform of $F(\omega)$
f_c	center frequency
G	direct current or steady field gradient
g	pulsed field gradient
h	microscopic r.f. field
h	Planck's constant
$\hbar$	Planck's constant divided by 2π
$\mathbf{h}_o$	incremental direct current magnetic field
h_x, h_y, h_z	components of **h**
$\mathcal{H}_Q$,	Hamiltonian operator for quadrupole interaction
$\mathcal{H}_{csa}$	Hamiltonian operator for chemical shift anisotropy
$\mathcal{H}_{sr}$,	Hamiltonian operator for spin rotation
$\mathcal{H}_d$	Hamiltonian operator for dipole-dipole
I	spin quantum number, also moment of inertia
i	unit vector along x axis
i	square root of minus one, also an index
J_{AX}	A-X spin-coupling constant in hertz
$J_i(\omega)$	spectral density function
J	angular momentum
J	rotational quantum number
j	unit vector in y direction
K	1024--referring to computer memory size
$K_i(r)$	correlation function
k	Boltzmann constant
k	unit vector in z direction
L	inductance
$\mathcal{J}$	torque
M	macroscopic magnetization vector
M_o	value of **M** at equilibrium

M_x, M_y, M_z	components of **M**
M	magnitude of **M**, also nuclear mass
m_i	magnetization due to nucleus i
O	a general operator
P	power
P_a	fractional population of site a
p	angular momentum vector
Q	quadrupole coupling tensor
Q_{ij}	components of **Q**
Q	quality factor of tuned r.f. circuit
R	resistance, also spectral resolution
R_1	spin-lattice relaxation rate ($R_1 = 1/T_1$)
R_2	spin-spin relaxation rate ($R_2 = 1/T_2$)
R_2^0	R_2 in the absence of scalar relaxation
$R_1(I)$	R_1 of nucleus I
$R_2(I)$	R_2 of nucleus I
$R_1(S)$	R_1 of nucleus S
$R_2(S)$	R_2 of nucleus S
$R(\tau)$	cross-correlation function
r	coordinate in spherical coordinate system
S	spin quantum number of nucleus S
S/N	signal-to-noise ratio
T_1	longitudinal or spin-lattice relaxation time
T_2	transverse or spin-spin relaxation time
T_2^*	time constant for decay of the free induction decay signal in presence of magnetic field inhomogeneity
$T_1(I)$	T_1 of nucleus I
$T_2(I)$	T_2 of nucleus I
$T_1(S)$	T_1 of nucleus S
$T_2(S)$	T_2 of nucleus S
T_r	rise or fall time of a signal
$T_{1\rho}$	T_1 in the rotating frame
$T_{2\rho}$	T_2 in the rotating frame

T_2^0	T_2 in the absence of scalar relaxation
t_p	width or time duration of r.f. pulse
t_{cp}	time between 180° pulses in a Carr-Purcell experiment
T_{ij}	components of the tensor, **T**
t	time
T	a tensor of the second rank
u	u mode, or dispersion signal
V	volume, also molecular rotation frequency
v	v mode or absorption frequency
v	frequency of r.f. radiation
W	power
W_α, W_β	spontaneous transition probability from α to β
W_+, W_-	transition probability with r.f. field
x	x axis in the laboratory frame
Y_i	spherical harmonic of order i
α	spin state for nucleus aligned parallel to B_o
β	spin state for nucleus aligned antiparallel to B_o
γ_I	gyromagnetic ratio for nucleus I
δ_I	chemical shift anisotropy for nucleus I
$\Delta 1/2$	line width at half maximum height
η	asymmetry parameter
$\hbar$	Planck's constant divided by 2π
$\sigma_\parallel$	chemical shift when internuclear vector is parallel to B_o
$\sigma_\perp$	chemical shift when internuclear vector is perpendicular to B_o
Σ_{ii}	components of the chemical shift tensor
τ_c	correlation time
χ	quadrupole coupling constant = e^2qQ/h
ω_o	Larmor precessional frequency in radians

ABBREVIATIONS

ADC	analog-to-digital converter
a.f.	audio frequency
COSY	(homonuclear) correlated spectroscopy
CP	Carr-Purcell T_2 experiment
CPMG	Carr-Purcell-Meiboom-Gill T_2 experiment, sometimes also called the PSCP or phase-shifted Carr-Purcell experiment
CSA	chemical shift anisotropy, $\sigma_{\parallel} - \sigma_{\perp}$
cw	continuous wave
dc	direct current
DEFT	driven equilibrium Fourier transform
DEPT	distortionless enhancement via polarization transfer
DFT	discrete Fourier transform
ERNST	enhancement via refocused nuclear spin-polarization transfer
FFT	fast Fourier transform (Cooley-Tukey algorithm)
FID	free induction decay (signal)
FT	Fourier transform

IC	integrated circuit
i.f.	intermediate frequency (usually 10.7 MHz)
HETCOR	heteronuclear correlated spectroscopy
INDOR	internuclear double resonance experiment
INEPT	insensitive nuclei enhanced via polarization transfer
ppm	parts per million
PSD	phase sensitive detector
r.f.	radio frequency
NMR	nuclear magnetic resonance
NOE	nuclear Overhauser enhancement
NOF	nuclear Overhauser factor
NQR	nuclear quadrupole resonance
2D-FT	two dimensional Fourier transform

CHAPTER ONE

THE FUNDAMENTAL CONCEPTS OF NUCLEAR MAGNETIC RESONANCE

Most organic and inorganic compounds which a chemist encounters exhibit a nuclear magnetism. In the absence of a large external magnetic field, the compounds are not magnetic. However, if they are placed in a large magnetic field, they will develop, in a short period of time, a gross macroscopic magnetic moment. This magnetism arises from the magnetic moments of the nuclei in the compound and is called nuclear magnetism. Such nuclear magnetism was first demonstrated by Lasarew and Schubnikow [1] for solid hydrogen in 1937. Using conventional, classical methods, they obtained the value of the magnetic moment for hydrogen with an error of about 10%. They also discovered that the microscopic nuclear magnets reached thermal equilibrium in a very short time period, even at the temperature of their experiment, which was about 2° K.

In the absence of a magnetic field, the nuclear magnets have no preferred orientation. In the presence of a large dc magnetic field, however, the nuclei prefer to be aligned with the magnetic field rather than against it. Thus, there are two different energy levels available to these microscopic magnets. In order for there to be a net macroscopic magnetic moment for the sample, there must be more of the nuclear magnets either in the lower energy level (in which the nuclear magnetic moments are aligned with the field) or in the upper energy level (in which the nuclear magnetic moments are aligned against the field).* In other words, the populations of the two energy levels must be different. In order to achieve this population difference between the two energy levels, the nuclei must give up some energy to the sample. Once the population difference is established it should be possible to equalize the populations by irradiation of the sample with energy of the correct

* An inverted magnetization which corresponds to more spins in the upper level can, of course, also give a measurable magnetization.

frequency. This irradiation should be accompanied by a net absorption of energy by the nuclei. In 1936 Gorter (2) tried to detect this energy absorption by classical, calorimetric methods using a sample of lithium fluoride. Due to the long relaxation time of the sample and the very low heat capacity of the spin system, the experiment failed. But the basic idea of the experiment was sound. In these experiments the relaxation time is a measure of the time required for the sample to reach thermal equilibrium. That is, it is the time required to go from no net magnetization in the absence of a magnetic field to the full net macroscopic magnetization (resulting from a net equilibrium difference between the two energy levels) in the presence of a magnetic field. The relaxation time, T_2, is also related to the spectral line width by the following equation, which is valid for Lorentzian lines,

$$g(\nu) = \frac{2\,T_2}{1 + 4\pi^2 T_2^2(\nu - \nu_o)^2}, \tag{1.01}$$

where $g(\nu)$ is the amplitude of the line at frequency ν and ν_o is the center frequency of the line. For such a Lorentzian line the full width of the line at half height is given by

$$\Delta_{1/2} = 1/\pi T_2. \tag{1.02}$$

A much better and more sensitive method for studying the properties of nuclear magnetization is the method of nuclear magnetic resonance (NMR). The first successful NMR experiments using garden-variety materials were done independently by Purcell, Torrey and Pound (3) and by Bloch, Hansen and Packard (4). The methods used by the two groups were quite different. Purcell and his coworkers measured the resonance absorption for the protons in solid paraffin. They detected the absorption of energy by the sample by measuring the additional loss which it caused in a tuned radio frequency (r.f.) bridge circuit at the resonance frequency. Bloch and his colleagues used a sample of water and detected the signal induced into a tuned r.f. circuit by the reorientation of the nuclear magnetic moments in the sample.

This early work by the Purcell group and by the Bloch group showed that nuclear magnetic resonance in bulk materials can be observed in several ways. The <u>slow passage</u> experiment consists of slowly sweeping the r.f. applied to a sample in a fixed magnetic field (or, alternatively, slowly sweeping the magnetic field with a fixed r.f.). The resonant signal is observed as a line of approximately Lorentzian shape. High-resolution NMR studies, and some broad line experiments as well, are carried out under conditions that approximate the requirements for slow passage. Adiabatic rapid passage requires that the sweep rate of either frequency or magnetic field be between certain limits. Adiabatic rapid passage is now used only infrequently in chemical applications--principally in the search for resonant frequencies of nuclei that give weak signals. The slow passage and rapid passage methods are usually referred to as continuous wave (cw) techniques, since the r.f. is applied continuously while the spectrum is observed.

A third method of observation put into practice initially by Hahn (5) makes use of short bursts, or pulses, of r.f. power at a discrete frequency. The observation of the nuclear spin system is made after the r.f. is turned off. Such techniques are called pulse methods or free precession techniques. Combination methods, in which r.f. pulses are applied but the transient behavior of the spin system is studied while the r.f. pulse is on, have also been developed (6).

In this book we describe the basic phenomena of the various pulse methods, the experimental techniques needed for their study, and some of the more important chemical applications of the methods. Although pulse methods were introduced almost as early as cw methods and have been improved in sophistication and versatility, they attracted the attention of chemists much later than the cw methods. The first person to really appreciate the power of the Fourier transform method and to apply it to real samples was Professor Ernst (7). In about 1970, with the advent of more sophisticated r.f. and digital electronics, people who in the normal course of their research did very little high-resolution NMR work discovered that carbon-13 NMR spectra could be done in reasonable time periods and with excellent results (8, 9).

Unfortunately, the free induction decay (FID), or time response signal, is not very easy to interpret. However, by executing a Fourier transformation of the FID data, a normal frequency spectrum can be obtained. In the past ten years these pulse NMR experiments have developed greatly in terms of the kinds of information available and in the ease of carrying out the experiments. It is now well known that more information is contained in the FID signal than in the cw spectrum. The FID signal obtained in the pulse NMR experiment can be obtained much more rapidly, and the pulse NMR technique has a versatility undreamed of in the conventional cw experiment. We shall discuss these experiments in great detail in Chapters 5, 6, and 7, and give a number of useful examples.

As we shall see in Chapter 2, nuclear relaxation plays an important role in pulse experiments, and pulse techniques provide the most generally useful method of measuring relaxation times. In the past, relaxation has often been dismissed as of no real importance in chemical problems, but with the advent of pulse methods that permit measurement of relaxation times for individual NMR lines in complex molecules, a whole new dimension is opened that complements measurements of chemical shifts and coupling constants in providing information about molecular structure and molecular dynamics. In Chapters 2 and 5 we describe these pulse methods and give examples of their use, while in Chapter 4 we try to provide some insight into the origin and mechanisms of relaxation. Relaxation times, as measured by pulse methods, can provide direct information on such important processes as chemical exchange, conformational changes in macromolecules, binding of small molecules to polymers, and molecular diffusion. In addition, relaxation time measurements can provide information about bond distances, the absolute sign of spin-spin coupling constants, the sign and magnitude of chemical shift anisotropies, and molecular correlation times. In Chapters 6 and 7 we discuss methods for making such investigations over a wide range of experimental conditions, together with examples of their uses.

In this introductory chapter we review some basic NMR properties and attempt to provide a conceptual framework for visualizing how nuclear magnets behave in pulse experiments. Of fundamental importance to our entire treatment in this book are two simple mathematical procedures: (a) the use of a rotating coordinate system to simplify the equations of motion of precessing nuclei, and (b) the use of Fourier transformation to establish the relation between the time course of certain processes and the basic frequencies characterizing those processes. We review the necessary background for these topics in Sections 1.1 and 1.5. Examples of their use occur continually throughout the book.

1.1 Relaxation and Nuclear Magnetic Energy Levels

It is well known from basic NMR theory (10) that an isolated nucleus of spin I, has 2I+1 energy levels, equally spaced with a separation, ΔE, given by

$$\Delta E = \mu B_o/I, \tag{1.1}$$

where B_o is the applied magnetic field and μ, the nuclear magnetic moment, is given by

$$\mu = \gamma hI/2\pi = \gamma\hbar I \tag{1.2}$$

Here γ is the gyromagnetic ratio, a constant for a given nucleus, h is Planck's constant, and $\hbar = h/2\pi$. From the usual Bohr relation, the frequency of radiation that induces a transition between adjacent levels is

$$\nu_o = \Delta E/h = \gamma B_o/2\pi \quad \text{Hz} \tag{1.3}$$

or

$$\omega_o = 2\pi\nu_o = \gamma B_o \quad \text{(rad/sec)}. \tag{1.4}$$

At equilibrium, nuclei are distributed among the energy levels according to a Boltzmann distribution. Following any process that disrupts this distribution, the nuclei will relax back to the equilibrium configuration. For example, if the sample is moved into or out of the magnetic field, or if the sample absorbs r.f. energy, the nuclear spin system returns to equilibrium with its surroundings, "the lattice," by a first-order relaxation process characterized by a time, T_1, the _spin-lattice relaxation time_. We shall discuss relaxation processes in Chapter 4. For the present it is sufficient to note that T_1 may vary over a wide range (approximately 10^{-4} - 10^{+4} sec). For small diamagnetic systems T_1 is typically of the order of 0.1 to about 10 sec, for a spin 1/2 nucleus.

Since a nucleus remains in a given energy level no longer than T_1 on the average, we can estimate the minimum width of the NMR line from the Heisenberg uncertainty principle:

$$\Delta E \Delta t \geq h, \tag{1.5}$$

where ΔE and Δt are the uncertainties in measurement of energy and time, respectively. For the case we are considering, with a line of full width at half-maximum (lwhm) intensity $\Delta_{1/2}$, this becomes

$$(h\Delta_{1/2})T_1 \geq h, \qquad \Delta_{1/2} \geq 1/T_1. \tag{1.6}$$

Often lines are broader than indicated by equation 1.6. For example, in solids or slowly tumbling molecules in liquids, magnetic dipole-dipole interactions between nuclei lead to appreciable line broadening, sometimes several kilohertz, as compared with the width of about 1 Hz expected from equation 1.6 for $T_1 = 1$ sec. To account for processes that cause the nuclear spins to come to equilibrium with each other, a second time, T_2, the spin-spin relaxation time, is defined so that

$$\Delta_{1/2} \approx 1/T_2. \tag{1.7}$$

To agree with the more precise definition of T_2 that we shall introduce in Section 1.2, we relate T_2 to the width of a Lorentzian line by the equation

$$\Delta_{1/2} = 1/(\pi T_2). \tag{1.8}$$

In this discussion we have been referring to the "natural" width of the resonance line determined by molecular processes. If the magnetic field is not perfectly homogeneous, nuclei in different parts of the sample experience slightly different values of the field; hence by equation 1.3 they resonate at slightly different frequencies. This leads to a contribution to the line width due to inhomogeneity (ΔB_0) of

$$\Delta_{1/2}(\text{inhom}) = \gamma \Delta B_0/(2\pi). \tag{1.9}$$

By analogy to equation 1.8, we can define a time T_2^* in terms of the observed line width as

$$\Delta_{1/2}(\text{obsvd}) = 1/(\pi T_2^*). \tag{1.10}$$

Thus T_2^* includes contributions from both natural line width and magnetic field inhomogeneity:

$$1/T_2^* = 1/T_2 + (\gamma \Delta B_0/2). \tag{1.11}$$

1.2 Macroscopic Magnetization and Demagnetization

Although many features of NMR spectra can be understood only by quantum mechanical considerations based on the energy level approach of Section 1.1, a number of properties are more easily visualized through a classical treatment. Most of the pulse experiments that we shall consider fall into this category. Hence we shall use the classical treatment almost exclusively. For this treatment we shall construct several simple physical pictures of the fundamental processes that take place in an NMR experiment. We shall use these simple concepts to design a number of useful experiments. The basic concepts that we shall discuss are:

1) nuclear magnetization processes
2) thermodynamic processes
3) motional processes of the magnetization vector, **M**

We shall use these basic concepts many times to discuss a number of the more common NMR experiments which we encounter in the following chapters.

We shall refer to nuclear magnetization processes as those processes which can be switched on and off very rapidly. For example, a magnetization vector can have its orientation changed very rapidly, in a microsecond or less. Spin-coupling effects can be switched on or off almost instantly. Nuclear magnetic energy levels can be switched on or off almost instantly. When a sample of water is placed in a magnetic field, the two energy levels discussed earlier are realized instantaneously. Similarly, when a spin decoupler is switched on, the spin-coupling interaction almost instantly goes to zero.

In contrast to those processes which can be switched on or off rapidly are processes which we shall call thermodynamic processes. These processes take place in the NMR experiment and are quite similar to more familiar thermodynamic occurrences, such as the heating (or cooling) of a glass of water by dropping a hot (or cold) rock into it. An appreciable amount of time is required for the system to reach thermodynamic equilibrium. As we mentioned earlier, Professor Gorter tried to measure some of the magnetic properties of materials by measuring the heating of a sample absorbing r.f. energy in a magnetic field. These early experiments failed due to the fact that the samples used had very long relaxation times and the heat capacity of the nuclear spin system was very small compared to the heat capacity of the lattice. If, however, we choose a very large value for μ by using electron spins, which have a magnetic moment 2000 times larger than a typical nuclear spin, a large value for the magnetic field B_o, and a low absolute temperature T, then we can discuss a classic experiment, the adiabatic demagnetization experiment. Consider that the macroscopic magnetic moment of the sample (this is just the sum of all the microscopic moments) is given by

$$\mathbf{M} = (Q\,\mu\,N\,B_o)/kT, \tag{1.12}$$

where Q is a constant, N is the total number of spins in the sample, k is the Boltzmann constant and T is the temperature in degrees Kelvin. The probability N_+ that a particle will have an energy E_+ is given by the Boltzmann relation

$$N_+ = N\exp(E_+/kT). \tag{1.13}$$

Therefore, the difference in population between the two energy levels, $\Delta E = E_+ - E_-$, is given by

$$\Delta N = N_+ - N_- = N\,(\Delta E)/kT \simeq N\,\mu\,B_o/kT. \tag{1.14}$$

Let us consider a paramagnetic salt which we shall put into a liquid helium bath at 4.2° K. When the entire system has come to equilibrium at 4.2° K, let us suddenly immerse the sample into a very large magnetic field. We know that the Boltzmann equation given above (equation 1.14) must be met. But initially $\Delta N = 0$ and ΔE is a very large number. One way to reconcile this situation is to postulate that the temperature T, which is the instantaneous temperature of the electron spins, is very hot. That is, T is very, very large. Thermodynamically, the spins must be infinitely hot if equation 1.14 is to be valid. Consequently, helium will be boiled off before the system comes to equilibrium again at 4.2° K. At equilibrium, T is a very small number again and both μ and B_o are relatively large numbers. Consequently, ΔN is a large number. If we reverse our steps and now suddenly remove the sample from the large magnetic field B_o, then B_o becomes zero but ΔN is still quite large. In order for equation 1.14 to hold, T must be very small; in fact, T must go to zero. This is actually what happens, and the spins are forced to a very low temperature approaching absolute zero. When the entire system reaches equilibrium again, we find that the sample temperature is much less than 4.2 K.

The purpose of this discussion is to demonstrate that spin-relaxation processes are thermodynamic processes and that rather substantial amounts of time are usually required for equilibrium to be attained. The reason for this delay is that the spins are only loosely coupled to the lattice. In other words, the spins do not interact very strongly with their surroundings. The same sorts of processes take place in nuclear systems, but due to the very low heat capacity of the nuclear spins, normal thermodynamic or calorimetric experiments cannot be successfully used in most cases. Thus far we have said nothing about how long it takes a sample to reach equilibrium after being placed in a magnetic field. Let us place a sample containing a milliliter of degassed benzene* in a magnetic field. In the presence of this field the hydrogen nuclei can be in one of two possible energy levels, as shown in Figure 1.1 below.

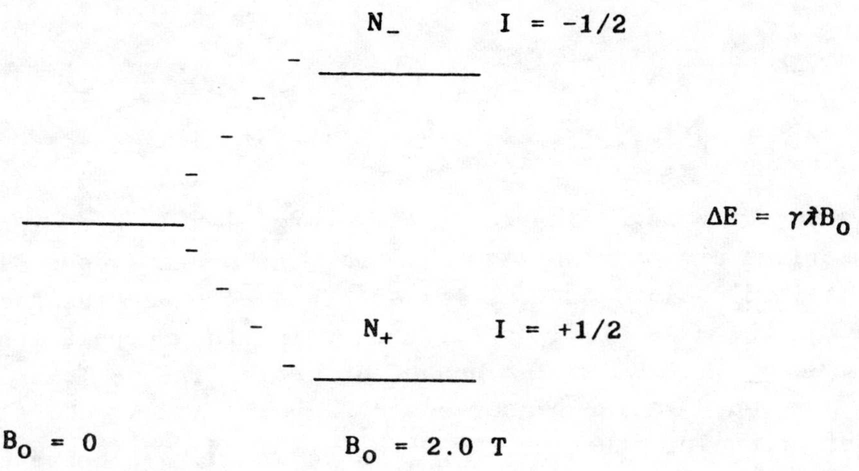

Figure 1.1 The energy level diagram for a single spin 1/2 system. In the absence of a dc magnetic field the levels are degenerate. For a field B_0, there are two energy levels whose difference ΔE is $\gamma\hbar B_0$.

* Oxygen gas is paramagnetic and affects the relaxation time of the benzene; therefore it is removed.

In the $I = +1/2$ state the nuclei are aligned with a component of their magnetic moments along the direction of the magnetic field, B_o. In the $I = -1/2$ state their moments are aligned antiparallel to B_o and they are in a high-energy state. As we have seen above (equation 1.13), according to the Boltzmann distribution there are, at equilibrium, more particles in the lower energy state than in the higher energy state. We would like to know a number of things about this energy distribution:

1) What is the population in each of the two states in the absence of the field?

2) What are the populations in the field at equilibrium?

3) How much time is required to reach equilibrium?

4) What happens if we irradiate the sample with r.f. energy?

5) How are the relative populations related to the NMR experiment?

In the absence of a magnetic field, the energy levels are degenerate and their populations are equal at equilibrium. As we shall see in the experiment described below, at the instant that we place the sample in the magnetic field there is still no difference in the populations. At equilibrium, however, if N_+ is the number of nuclei in the lower energy level, N_- is the number in the higher energy level, and $N = N_+ + N_-$ is the total number of nuclei, then at $t = 0$, $N_+ = N_- = N/2$. At equilibrium we know from Boltzmann's law that

$$\frac{N_+^o}{N_-^o} = \exp\left[\frac{\Delta E}{kT}\right] = \exp\left[\frac{\gamma h B_o}{2\pi kT}\right] = \exp\left[\frac{2\mu B_o}{kT}\right], \qquad (1.15)$$

where

N_+ = instantaneous population of the I = + 1/2 state
N_+^o = equilibrium population of the I = + 1/2 state
N_- = instantaneous population of the I = - 1/2 state
N_-^o = equilibrium population of the I = - 1/2 state
N = $N_+ + N_-$ = the total number of nuclei
ΔE = difference between the energy levels
k = Boltzmann's constant
T = temperature in degrees Kelvin
γ = proton gyromagnetic ratio
h = Planck's constant
B_o = magnetic field strength in tesla
W_+ = probability for a transition from the +1/2 to the -1/2 state in the presence of an r.f. field
W_- = probability for a transition from the -1/2 to the +1/2 state in the presence of an r.f. field
n^o = equilibrium population difference
n = instantaneous population difference

In order to get some insight into how rapidly the sample reaches equilibrium, let us carry out a simple, but very useful, experiment. Let us place the benzene sample next to a small compass and measure the deflection of the compass needle as a function of the amount of time, x, the sample has spent in the magnetic field. We proceed as follows:

1) Initially, we set the time, x = 10 sec.

2) Place the sample in the field for x sec. Withdraw the sample and immediately measure the deflection. Record this value.

3) If the deflection is not substantially greater than the previous reading, go to step 6; otherwise proceed to step 4.

4) Set the sample aside in a place not near the magnetic field and leave it there for about 10 min (so that any magnetization built up can decay to zero).

5) Increase x by 10 sec and go to step 2.

6) Plot the results.

Such an experiment will give results such as those shown in Figure 1.2a.

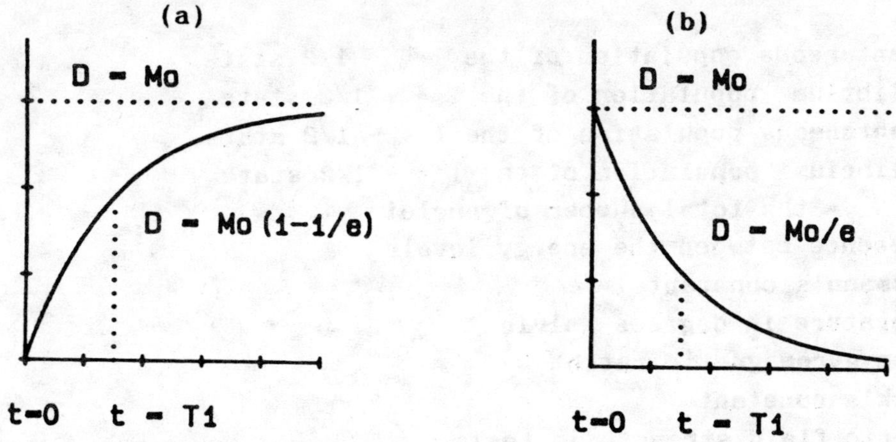

Figure 1.2 (a) An initially unmagnetized sample (e.g., water) will become magnetized at an exponential rate, up to a maximum value of M_0, if placed in a dc magnetic field. (b) The completely magnetized sample will lose its magnetization at an exponential rate if removed from the magnetic field.

In Figure 1.2a we see that the deflection increases exponentially to a maximum value. This indicates that the gross macroscopic magnetization increases exponentially to its equilibrium value, M_0. The time required for the the magnetization to reach $M_0[1-(1/e)]$ (where e = 2.71828...) is called the spin-lattice relaxation time and is designated with the symbol T_1. The value of T_1 can also be measured by letting the sample reach its full magnetization M_0 and then monitoring its decay back to zero magnetization in 10 sec intervals. The result of this experiment is shown in Figure 1.2b. As is expected, the magnetization decays to M_0/e in a time $t = T_1$. As we shall see later, the magnitude of M_0 is proportional to the equilibrium population difference, n_0, which in turn is proportional to the strength of the magnetic field, B_0, and the gyromagnetic ratio, γ.

From this experiment we know that there must be some mechanism which allows a population difference to take place. In order to see how this comes about, let us denote by W_α the probability of a spontaneous transition from the +1/2 state to the -1/2 state and by W_β the probability of a spontaneous transition from the -1/2 state to the +1/2 state. We can then write

$$dN_+/dt = \dot{N}_+ = N_-W_\beta - N_+W_\alpha. \qquad (1.16)$$

Clearly, $W_\alpha \neq W_\beta$, or we would never reach equilibrium. Furthermore, we know that at equilibrium $dN_+/dt = 0$, so

$$N_-^0 W_\beta - N_+^0 W_\alpha = 0$$

or

$$\frac{N^o_-}{N^o_+} = \frac{W_\alpha}{W_\beta} . \qquad (1.17)$$

But recall from Boltzmann's law (equation 1.15 above) that

$$\frac{N^o_-}{N^o_+} = \exp\left[\frac{-\gamma h B_o}{2\pi kT}\right] = \frac{W_\alpha}{W_\beta} , \qquad (1.18)$$

or

$$\frac{W_\beta}{W_\alpha} = \exp\left[\frac{\gamma h B_o}{2\pi kT}\right] . \qquad (1.19)$$

Now we can rewrite equation 1.16 in terms of the instaneous population difference, n, and the total population, N, as follows:

$$dn/dt = N(W_\beta - W_\alpha) - n(W_\beta + W_\alpha) . \qquad (1.20)$$

If we define n^o, R_1 and T_1 by

$$n^o = N\left[\frac{W_\beta - W_\alpha}{W_\beta + W_\alpha}\right] \qquad (1.21)$$

and

$$1/T_1 = R_1 = (W_\beta + W_\alpha) , \qquad (1.22)$$

then equation 1.20 may be written as

$$dn/dt = R_1(n^o - n), \qquad (1.23)$$

where n^o is the equilibrium population difference and R_1 is the spin-lattice relaxation rate. If we rearrange equation 1.23 we obtain

$$\frac{1}{(n^o - n)} dn = R_1 dt .$$

The solution to this equation is

$$- \log_e(n^o - n) = R_1 t + A ,$$

or

$$n^o - n = Ae^{-R_1 t} = A \exp(-R_1 t) . \qquad (1.24)$$

If n=0 at time t=0, that is, if the sample is initially unmagnetized, then $A = n^o$ and

$$n = n^o [1 - \exp(-R_1 t)] . \qquad (1.25)$$

A plot of n as a function of time gives a result identical to that obtained in Figure 1.2a. If the sample is initially fully magnetized then

$$n = n^o \exp(-R_1 t) . \qquad (1.26)$$

In this case a plot of n versus time gives the same result as that shown in Figure 1.2b.

These simple experiments tell us that our sample will reach equilibrium according to the first-order differential equation (1.23) at a characteristic rate R_1, or with a characteristic time T_1. We will later on look in more detail at the factors which determine the magnitude of T_1. At this point we have observed that by some mechanism the sample will reach an equilibrium configuration in which a Boltzmann population of the energy levels is realized. We now ask what happens when we turn on the r.f. transmitter in our NMR spectrometer and begin to irradiate the sample at a frequency $v = \Delta E/h$. Here, as in most forms of spectroscopy, the probability of a stimulated transition, W_-, from the upper state to the lower state (-1/2 to +1/2) and the probability of a stimulated transition, W_+, from the lower to the upper level (+1/2 to -1/2) are equal; thus

$$W_- = W_+ = W . \qquad (1.27)$$

We might well ask at this point why the probabilities here are the same while we maintained just above that $W_\alpha \neq W_\beta$. In the previous case there was no external r.f. radiation present. For the sample to reach equilibrium while interacting only with the "lattice" (the thermal motions of the molecules), it is necessary that the lattice be in an energy state which can give up or receive energy. The fact that the spins are initially in a "hot" thermodynamic state compared to the lattice means that there is a much higher probability that the lattice will receive energy from the spins than vice versa.

In the presence of the external r.f. field from the amplifier in the spectrometer, $W_- = W_+ = W$. Since at equilibrium there are more spins in the lower energy level, there will be a net absorption of energy by the sample. The rate of energy absorption, which is proportional to the magnitude of the signal which we will observe, is proportional to the rate of change of population of the lower energy level, $dN_+/dt = N_+$:

$$dN_+/dt = W(N_- - N_+). \qquad (1.28)$$

Since $N = N_+ + N_-$ and $n = N_+ - N_-$,

then
$$N_+ = 1/2\ (N + n)\ , \qquad (1.29)$$
$$N_- = 1/2\ (N - n)\ .$$

If we substitute (1.28) into (1.27) we obtain

$$dn/dt = -2Wn\ . \qquad (1.30)$$

This equation is also a first-order differential equation and has the solution

$$n = n(0)\ e^{-2Wt}\ , \qquad (1.31)$$

where n(0) is the value of n at the time that we turn on the r.f. field. Equation 1.30 tells us that our population difference, n, will disappear under the influence of the r.f. irradiation if we neglect the effects of relaxation (we consider both effects together below). This point is important to remember because, as mentioned above, the magnitude of the signal is proportional to W, the irradiation power. Consider that the signal intensity is proportional to the rate of change of energy absorption, dE/dt, which is given by

$$dE/dt = N_+Wh\omega - N_-Wh\omega = h\omega Wn , \qquad (1.32)$$

where ω is the frequency of the irradiation. The faster we drive spins from the lower level to the upper level, the greater the signal, but also the faster we drive the sample into a state where the populations are again equal and the signal disappears. This we often refer to as a state of sample "saturation."

Thus we see that in order to see a signal we must have a net absorption of energy and this requires that the population difference, n, must be greater than zero. What we must do, of course, is combine equations 1.30 and 1.23 in order to account for the two competing processes

$$dn/dt = -2Wn + (n - n^o)R_1 . \qquad (1.33)$$

Thermodynamics is at work trying to maintain a Boltzmann distribution between the energy levels and an external r.f. field is trying to equalize the same populations. In the steady state dn/dt is zero so

$$n = \frac{n^o}{1 + 2WT_1} . \qquad (1.34)$$

We will obtain the maximum signal-to-noise ratio (S/N) at the point where $2WT_1 \approx 1$. As long as $2WT_1 \ll 1$, $n \approx n^o$ and irradiating the sample does not appreciably disturb our thermal equilibrium.

The main points to remember are:

1) A net macroscopic magnetization will develop in most organic samples due to thermodynamic relaxation processes.

2) These thermodynamic processes require time, usually on the order of seconds to minutes, to reach equilibrium, that is, to develop a macroscopic magnetization.

3) Other processes are present which can act almost instantaneously. The different energy levels are realized the instant that the sample is placed in the field. A r.f. field can almost instantly equalize the populations of the energy levels.

In summary, most commonly encountered organic samples are diamagnetic and, when placed in a strong magnetic field, will themselves develop a macroscopic magnetic moment along the field direction. Attaining this equilibrium magnetization is a thermodynamic process and, as such, requires time. The time required to attain equilibrium can usually be described by a parameter called the spin-lattice relaxation time, T_1. (T_1 will be discussed in more detail in Chapter 4.) In order to simplify our discussions of T_1 and other relaxation parameters, we shall first introduce some elementary properties of angular momentum and the concept of the rotating frame of reference.

1.3 Angular Momentum and Magnetic Moments in the Rotating Frame

Consider a small gyroscopic top. It has angular momentum associated with it which is a function of how fast it is spinning, its size and the magnitude of the mass which is spinning. We define a number of terms to simplify the discussion:

v = the linear velocity of the spinning mass
m = the rotating mass of the top
r = the radius of gyration of the top
p = the linear momentum of the top
J = the angular momentum of the top
L = the torque exerted on the top
F = the force exerted on the spinning mass of the top

These various parameters are shown schematically in Figure 1.3.

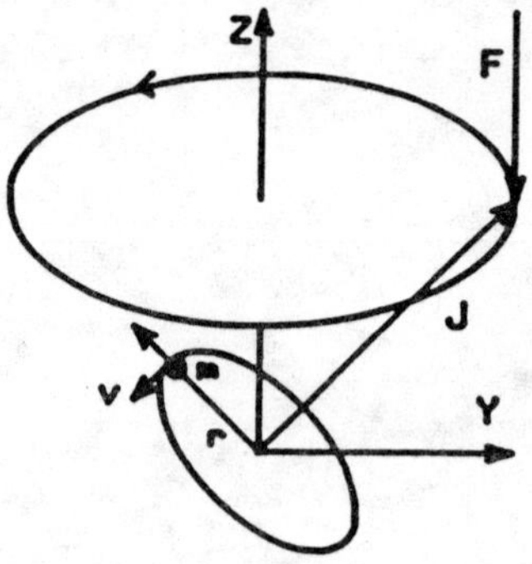

Figure 1.3 If a force F is imposed on a body with angular momentum, J, it will precess about the axis of the force at a rate dJ/dt = L = r x F.

A short summary of vector notation and the vector dot and cross products is given in Appendix A. This summary will be helpful in the following discussion and others throughout the book.

The total angular momentum is related to the linear momentum by the equation

$$\mathbf{J} = \mathbf{r} \times \mathbf{p} = \mathbf{r} \times (m\mathbf{v}) = m\mathbf{r} \times \mathbf{v}, \tag{1.35}$$

and the time rate of change of J, dJ/dt, is given by

$$dJ/dt = d(\mathbf{r} \times \mathbf{p})/dt$$
$$= dr/dt \times \mathbf{p} + \mathbf{r} \times dp/dt. \qquad (1.36)$$

However,

$$dr/dt \times \mathbf{p} = \mathbf{v} \times \mathbf{p} = m\mathbf{v} \times \mathbf{v} = 0, \qquad (1.37)$$

and by Newton's second law of motion

$$\mathbf{r} \times dp/dt = \mathbf{r} \times \mathbf{F} = \mathbf{L}; \qquad (1.38)$$

hence,

$$dJ/dt = \mathbf{L}. \qquad (1.39)$$

That is, the greater the torque **L** on the top, the faster it will precess. It is important to note here that it is usually the case that it is the direction of the angular momentum that changes and not the magnitude. In the case of our small gyroscopic top, the magnitude of the angular momentum remains constant. If we exert a torque **L** on it, it will vary in time according to equation 1.39 above, but it is only the direction of J that changes. In other words, the top precesses, or wobbles, about the axis along which we apply the torque (see Figure 1.4).

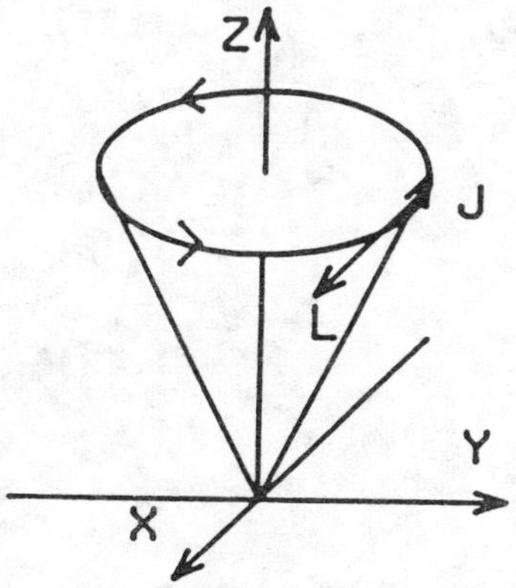

Figure 1.4 An object with angular momentum J will precess if a torque L is applied to it, dJ/dt = L.

In NMR experiments, we deal with nuclei such as protons which possess angular momentum and thus are very similar to the gyroscopic top which we discussed above. Of course, in the case of protons, the magnitude of J is very small indeed. Protons possess another property which is very important. They have a nuclear magnetic moment. Thus, if a proton is placed in a magnetic field it experiences a torque due to the interaction between its magnetic moment and the field of the magnet. The larger the value of the magnetic field B_0, the faster the rate of precession. This is the classical explanation for the fact that the resonance frequency for a particular nucleus increases as the magnetic field intensity increases. For each of the nuclear magnets with which we shall be working there is a precise relation between its intrinsic angular momentum J, which is a fundamental physical constant, and its magnetic moment μ, which is also a fundamental constant. This relation is stated mathematically by

$$\mu = \gamma J, \tag{1.40}$$

where γ is a constant for a particular nucleus and is called the gyromagnetic ratio (constant), since it is the ratio of the magnetic moment of the nucleus to the angular momentum of the nucleus.

Let us consider a simple experiment. Place a sample of water in a magnetic field of, say, 2.3 tesla (23,000 gauss). Due to the interaction between the magnetic moment of the protons and the magnetic field, the protons will experience a torque L, which is given by

$$L = \mu \times B_0. \tag{1.41}$$

This torque will cause the protons to precess at a rate given by

$$dJ/dt = L = \mu \times B_0 \tag{1.42}$$

or

$$d\mu/dt = \gamma\, dJ/dt = \gamma(\mu \times B_0) \tag{1.43}$$

(see Figure 1.5).

Equation 1.43 tells us that the rate at which nuclei precess, or resonate, in a magnetic field depends upon the magnitude of μ and on the magnitude of B_0. Since the magnetic moment of a proton is four times greater than the magnetic moment for a carbon-13 nucleus, the proton resonance frequency will be four times greater than the carbon-13 resonance frequency for any given field strength.

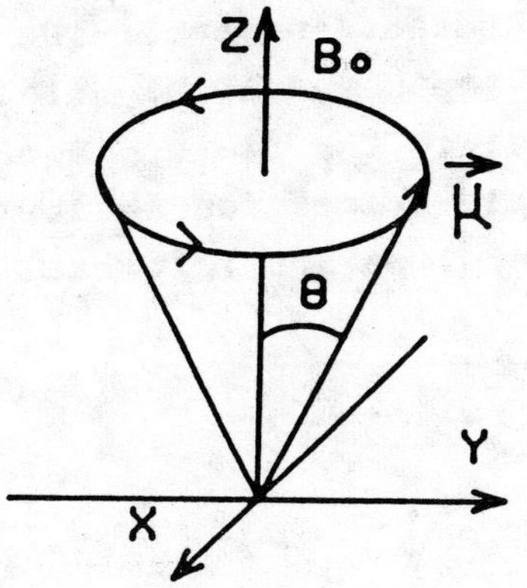

Figure 1.5 A particle with angular momentum J and a magnetic moment μ will precess about a magnetic field $B_o = kB_z$. The magnetic moment precesses at a rate proportional to μ and B_o. $d\mu/dt = \gamma dJdt = \gamma(\mu \times B_o) = \omega_o \times \mu$.

In any particular NMR experiment the magnitudes of both μ and B_o are constant; consequently the macroscopic magnetic moment will precess at a constant frequency about the direction of the magnetic field (see Figure 1.5 above). This can be expressed as

$$d\mu/dt = \omega_o \times \mu = -\mu \times \omega_o = +\gamma(\mu \times B_o). \qquad (1.44)$$

Hence,

$$\omega_o = -\gamma B_o \quad \text{(rad/sec)} \qquad (1.45)$$

or

$$v_o = \omega_o/2\pi = -\gamma B_o/2\pi \quad \text{(Hz)}. \qquad (1.46)$$

Equation 1.45 (or 1.46) is <u>the</u> fundamental NMR equation and is given a special name, the <u>Larmor equation</u>. If we know the value of γ for the proton (4.2576 kHz/gauss), then we can calculate the resonance frequency for any given magnetic field strength. The negative sign means that the motion is in the direction indicated in Figure 1.5a if $\gamma > 0$.

In most of the physical pictures we develop in this book, we would like to keep our model as simple as possible. This is difficult to do if we must continually be concerned with the motions of all of the individual microscopic magnetic moments and if we must be concerned at all times with the precession of the nuclei about B_o. For this reason we shall resort to two very helpful simplifications (the simplifications are realistic since in any event one cannot observe the individual microscopic moments). Rather than depict the motions of the nuclei as in Figure 1.6a, we shall use the simplification shown in Figure 1.6b.

That is, rather than be concerned with an almost equal number of nuclei precessing parallel and anti parallel to B_0 (which we shall always arrange to be along the z axis in our coordinate system), we shall be concerned only with the net, average macroscopic moment **M**. Since there are slightly more nuclei oriented parallel to the field, **M** will point along the +z axis at equilibrium. Since rapid precession about z causes the x and y components of **M** to average to zero, we are left with only a z component of **M** (see Figure 1.6b).

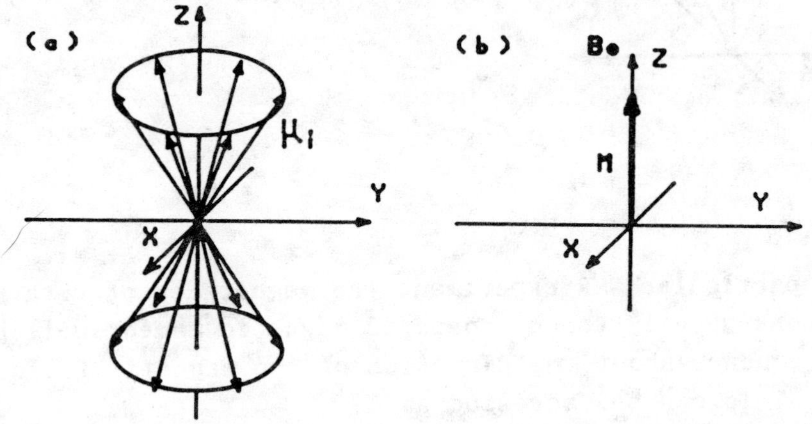

Figure 1.6 (a) An ensemble of spin 1/2 nuclei (μ_I) in a magnetic field will precess about the direction of the magnetic field. (b) The net macroscopic magnetization, **M**, is oriented along the direction of the magnetic field, B_0.

We shall simplify our model further by using a rotating frame of reference. Thus we refer the motion of the magnetization, not to the fixed coordinate system of the laboratory, but to a coordinate system that rotates about B_0 in the same direction in which the nuclear moments precess. This we shall refer to as the <u>rotating frame</u>.

The concept of a rotating coordinate system should be familiar, since we all customarily refer our position and motion relative to the earth, a coordinate system that rotates at the rate of $2\pi/24$ rad/hr. A man standing "at rest" on the equator would appear to a distant observer to be moving with a velocity of over 1000 miles/hr. And if our subject repeatedly threw a ball "vertically" upward and allowed it to fall in the earth's gravitational field, the ball would appear to him to describe a simple vertical straight line path and to be subject to no horizontal forces. To the distant observer, however, the ball would traverse a complicated path.

In a similar way, the magnetization vector describes a much simpler path in a suitable rotating frame than in the laboratory frame. In order to treat the system quantitatively we must see how the basic equation of motion of **M** is altered in the rotating frame. Generally in this book we refrain from detailed derivations, but in this case the derivation seems to be warranted, both because the result is very important to our later discussion and because the result, while simple, is not intuitively obvious.

To derive the basic equation of the rotating frame, we shall first develop a simple and well-known result for the time derivative of a vector **M** in terms of its components. Let

$$\mathbf{M} = \mathbf{i}\, M_x + \mathbf{j}\, M_y + \mathbf{k}\, M_z. \qquad (1.47)$$

With the usual expression for the differentiation of products (16), then

$$d\mathbf{M}/dt = \mathbf{i}\; \partial M_x/\partial t + \mathbf{j}\; \partial M_y/\partial t + \mathbf{k}\; \partial M_z/\partial t$$

$$+ M_x\, \partial \mathbf{i}/\partial t \; + \; M_y\, \partial \mathbf{j}/\partial t \; + \; M_z\, \partial \mathbf{k}/\partial t. \qquad (1.48)$$

Since **i**, **j**, and **k** are <u>unit</u> vectors, their time derivatives cannot change their lengths but only their direction. Mathematically, rotation is described in terms of a vector cross-product (8):

$$\partial \mathbf{i}/\partial t = \boldsymbol{\omega} \times \mathbf{i}, \quad \partial \mathbf{j}/\partial t = \boldsymbol{\omega} \times \mathbf{j}, \quad \partial \mathbf{k}/\partial t = \boldsymbol{\omega} \times \mathbf{k}. \qquad (1.49)$$

The magnitude of $\boldsymbol{\omega}$ gives the angular frequency of rotation of the unit vector, while the direction of $\boldsymbol{\omega}$ is the axis about which the rotation occurs. In such a rotating coordinate system all three unit vectors rotate with a velocity and direction defined by the same $\boldsymbol{\omega}$; hence equation (1.48) becomes

$$(d\mathbf{M}/dt)_{fixed} = \partial \mathbf{M}/\partial t + \boldsymbol{\omega} \times (\mathbf{i}\, M_x + \mathbf{j}\, M_y + \mathbf{k}\, M_z)$$

$$= (\partial \mathbf{M}/\partial t)_{rot} + \boldsymbol{\omega} \times \mathbf{M}. \qquad (1.50)$$

The total derivative represents the overall motion of **M** in the fixed (laboratory) frame, whereas the partial derivative represents the explicit dependence of **M** on time in the rotating frame. (We shall subsequently use the ordinary, rather than the partial, derivative symbol, with a specific subscript, as in equation 1.50, when there is ambiguity.)

If **M** represents the magnetization vector, then from equation 1.44

$$(d\mathbf{M}/dt)_{fixed} = \gamma \mathbf{M} \times \mathbf{B} \qquad (1.51)$$

and from equation 1.50

$$(d\mathbf{M}/dt)_{rot} = \gamma \mathbf{M} \times \mathbf{B} - \boldsymbol{\omega} \times \mathbf{M}. \qquad (1.52)$$

Rearranging terms in equation 1.52, and making use of the fact that for two vectors **A** and **B**, $\mathbf{A} \times \mathbf{B} = -\mathbf{B} \times \mathbf{A}$, we obtain

$$(d\mathbf{M}/dt)_{rot} = \gamma\mathbf{M} \times \mathbf{B} + \gamma\mathbf{M} \times \boldsymbol{\omega}/\gamma$$

$$= \gamma\mathbf{M} \times (\mathbf{B} + \boldsymbol{\omega}/\gamma). \qquad (1.53)$$

The term $\boldsymbol{\omega}/\gamma$ has the dimensions of a magnetic field and can be considered a "fictitious" field that arises from the effect of the rotation. Alternatively, we can write equation 1.53 in terms of an <u>effective</u> field,

$$(d\mathbf{M}/dt)_{rot} = \gamma\mathbf{M} \times \mathbf{B}_{eff} \qquad (1.54)$$

where

$$\mathbf{B}_{eff} = \mathbf{B} + \boldsymbol{\omega}/\gamma. \qquad (1.55)$$

Equations 1.53 and 1.54 tell us that the ordinary equations of motion applicable in the laboratory frame are valid in the rotating frame as well, provided $\mathbf{B}_{eff}$ as defined in equation 1.55 is used in place of **B**. Thus, in the rotating frame, the magnetization precesses about $\mathbf{B}_{eff}$. Normally we shall be interested in a frame that rotates at or near the frequency of an applied r.f. field $\mathbf{B}_1$, since that choice of rotating frame usually results in the simplest treatment.

1.4 Magnetization in the Rotating Frame

With the aid of the rotating frame we can explain just how the magnetization behaves during certain types of NMR experiments and see more explicitly the effects of relaxation and of magnetic field inhomogeneity. In this section we try to rationalize some apparent anomalies found in comparing cw and pulse experiments.

First let us treat the case where the only magnetic field present is of magnitude B_o and, as usual is applied along the z axis; then $\mathbf{B} = \mathbf{B}_o$. Consider a frame rotating at an angular frequency $\boldsymbol{\omega}_o = -\gamma\mathbf{B}_o$, the Larmor frequency of equation 1.45. Under these conditions equation 1.55 reduces to $\mathbf{B}_{eff} = 0$, and from equation 1.35, **M** is invariant with time in this rotating frame. That is, **M** remains stationary along the z' axis in the rotating frame. This is really just a restatement of the Larmor equation in other terms.

Now suppose that in addition to $\mathbf{B}_o$ we also apply $\mathbf{B}_1$ perpendicular to $\mathbf{B}_o$ (i.e., in the xy plane), and $\mathbf{B}_1$ rotates in the laboratory frame at ω rad/sec. (Strictly speaking, $\mathbf{B}_1$ oscillates, but such an oscillating field can be treated as the sum of two fields rotating in opposite directions. Only the component rotating in phase with the nuclear precession has an effect on the magnetization vector.) Then in the frame rotating at ω, the effective field becomes

$$\mathbf{B}_{eff} = \mathbf{B}_o + \boldsymbol{\omega}/\gamma + \mathbf{B}_1. \qquad (1.56)$$

At resonance the fictitious field exactly cancels $\mathbf{B}_o$ along the z axis and leaves only $\mathbf{B}_1$ in the xy plane to interact with **M**. Since $\mathbf{B}_1$ rotates at the same frequency as the frame, we can arbitrarily assign B_1 along the rotating x axis, designated the x' axis. Equation 1.53 tells us then that <u>in the rotating frame</u> **M** precesses about the x' axis, as indicated in Figure 1.7a. From the Larmor equation the angular precession frequency about the x' axis is γB_1. Since γB_1 varies from about 0.1 milligauss for high resolution cw experiments to about 100 gauss for some pulse studies, the precession frequency for protons ranges from 3 to 3×10^6 rad/sec. In a time t_p (sec), the angle θ through which **M** precesses is given by

$$\theta = \gamma B_1 t_p \quad \text{(rad)}. \qquad (1.57)$$

As we shall see in Chapter 2, this relation is fundamental to the use of pulse methods. It is important to note that in the laboratory frame the motion of **M** is rather complex: a rapid precession about B_0 combined with a much slower precession about B_1.

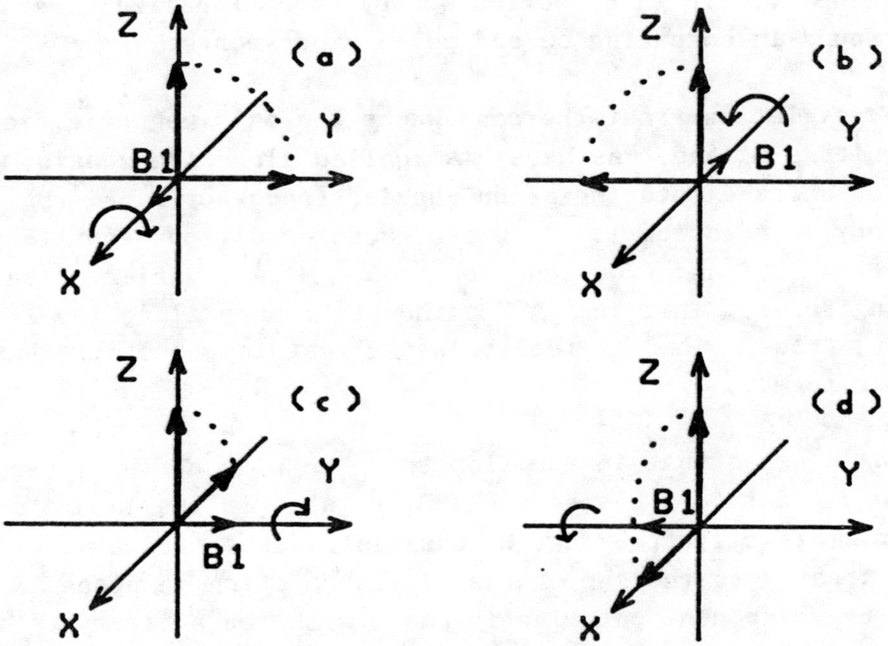

Figure 1.7 For 90° γB_1 pulses applied along the x, y, -x, and -y axes M will be rotated 90° about those axes and will lie along the y, -x, -y, and x axes, respectively. In (a), γB_1 is along the +x axis and M is rotated to +y. In (b), γB_1 is along the -x axis and M is rotated to -y. In (c), γB_1 is along the +y axis and M is rotated to -x. In (d), γB_1 is along along the -y axis and M is rotated to the +x axis..

We now inquire what happens if B_1 is applied along the x' axis long enough to cause **M** (remember that **M** is really an ensemble of nuclear moments with a small excess number of them oriented along the z' axis.) to tip through an angle of θ toward the y' axis. After B_1 is turned off, the situation is that shown in Figure 1.7a. Since there was originally no phase coherence of the moments in the x'y' plane, there was no component of **M** in this plane. Now, however, as a result of being tipped, the moments lie at such an orientation that a component of **M** is generated along y'. In this orientation the precessing moment is capable of inducing a r.f. voltage in a receiver coil placed along the y axis of the laboratory frame. Because of the natural processes that cause nuclei to exchange energy with each other, the moments begin to spread out in the x'y' plane, as indicated in Figure 1.7b. Thus, $M_{y'}$ decays with a time constant T_2. Since the magnetic field is not perfectly homogeneous, nuclei in different portions of the sample experience slightly different values of $\mathbf{B}_0$, and hence precess at slightly different frequencies, some faster than the rotating frame and some slower. This process, too, causes $M_{y'}$ to diminish; overall, $M_{y'}$ decreases to zero with a time constant T_2^*, as defined in equation 1.11. Spin-lattice relaxation also occurs as the nuclear moments lose energy to their surroundings and relax, or tip back, toward the z' axis, as indicated in Figure 1.7c. Thus $M_{z'}$ relaxes toward its equilibrium value of M_0 with time constant T_1. It is clear from Figure 1.7d that by the time the moments have returned to their original positions and thus restored $M_{z'}$ to a value of M_0, there can be no components of $\mathbf{M}$ in the x'y' plane. Therefore the time characterizing loss of $M_{y'}$, namely T_2 (or T_2^* if inhomogeneity effects are important), can never be longer than T_1, the time constant for restoration of $M_{z'}$ to its equilibrium value. On the other hand, T_2 (or T_2^*) can be shorter than T_1. In general, then,

$$T_2^* \leqslant T_2 \leqslant T_1. \qquad (1.58)$$

We shall often be concerned with the situation in which $\mathbf{B}_1$ and the rotating frame move at frequencies different from the resonance frequency. In general, then, we encounter the relation depicted in Figure 1.8, where the fictitious field does not cancel $\mathbf{B}_0$ and leaves a component of the effective field in the z' direction. Then

$$\begin{aligned} \mathbf{B}_{eff} &= [(B_0 - \omega/\gamma)^2 + B_1^2]^{1/2} \\ &= (1/\gamma)[(\gamma B_0 - \omega)^2 + (\gamma B_1)^2]^{1/2} \qquad (1.59) \\ &= (1/\gamma)[(\omega_0 - \omega)^2 + (\gamma B_1)^2]^{1/2}. \end{aligned}$$

In this case **M** precesses about B_{eff} in the rotating frame at a frequency given by

$$\Omega = [(\omega_o - \omega)^2 + (\gamma B_1)^2]^{1/2}. \qquad (1.60)$$

The off-resonance condition ($\omega_o - \omega \neq 0$) may arise from the presence of: (a) several chemically shifted nuclei, so that some or all precess at a frequency different from the r.f.; (b) magnetic field inhomogeneity, so that the precession frequencies of nuclei in different parts of the sample are different; or (c) static dipolar fields in solids, where each nuclear moment experiences not only the applied field but also the local field due to neighboring nuclei. We shall encounter each of these situations in Chapters 5 and 6.

With the use of the rotating frame we can clarify some important features of the pulse experiments. Consider first the cw experiment carried out by sweeping the magnetic field, B_o. If we take the frame rotating, as usual, at the frequency of B_1, then for B_o far above resonance the z' component of B_{eff} is large (see Figure 1.8) and $B_{eff} \approx B_o$. In this equilibrium situation **M** lies along B_o. As B_o decreases toward the resonance value, the component along z' in Figure 1.8 decreases and B_{eff} tips away from the z' axis. It has been shown (11, 12) that if the rate at which B_{eff} tips is slow enough, **M** is able to "follow" B_{eff}, that is, to remain continuously aligned with B_{eff}. If B_1 is along the x' axis, **M** is aligned along this axis at resonance, and as resonance is passed **M** moves on to become aligned eventually along the -z' axis. The requirement that B_{eff} tip slowly (11) is

$$dB_o/dt \ll \gamma B_1^2. \qquad (1.61)$$

Equation 1.61 is the adiabatic theorem, and a field sweep that conforms to this condition is known as adiabatic passage. In principle, the slow

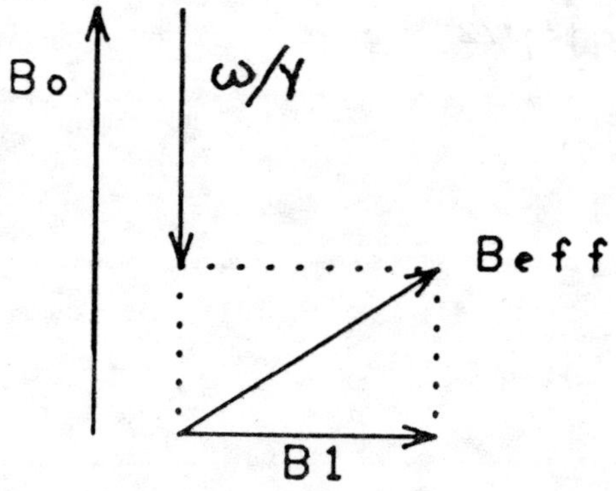

Figure 1.8 The effective magnetic field in the rotating frame, B_{eff}, is the vector sum of the applied field B_o, the "fictitious field" ω/γ, and the γB_1 field. During an r.f. pulse the magnetization, **M**, rotates about the effective field, B_{eff}.

passage used for high-resolution studies conforms to the adiabatic condition, but in practice, at the B_1 levels usually employed in cw work (approximately 0.1 milligauss), sweep rates much slower than 1 Hz/sec would be needed. Adiabatic passage, as the term is usually used, requires not only that the adiabatic theorem be obeyed but also that no relaxation occur during the sweep. Thus the complete condition for adiabatic passage is

$$1/T_2 \ll (1/B_1)dB_0/dt \ll \gamma B_1. \qquad (1.62)$$

(For solids $1/T_2$ can be replaced by $1/T_1$, which is a less stringent condition [11].) Because there is a minimum sweep rate, adiabatic passage is usually called "adiabatic rapid passage." Typical sweep rates are of the order of 10 milligauss/sec with $B_1 \approx 10$ milligauss (at least for carbon-13 experiments [13]). As we shall see in Chapter 6, adiabatic rapid passage can be used to invert the magnetization for certain experiments.

We must contrast the situation implied by the adiabatic theorem, which results in **M** always remaining aligned with B_{eff}, with that occurring when an r.f. pulse is applied to a spin system. In this case the system is initially at equilibrium at or near resonance, so that B_{eff}, which lies along z', is small or zero. A relatively large B_1 (often 10 to 100 gauss or more) is applied suddenly (typical rise times are 50 to 500 nsec). Thus B_{eff} is abruptly changed from the z' to the x' axis. **M** is unable to follow the change and, as we have seen, precesses in the field of B_{eff}, approximately in the y'z' plane.

Before continuing with more details of free induction decays and spin echoes, we should discuss, at least briefly, some points which have been a source of confusion to students who are just beginning their work in pulse NMR spectroscopy. From quantum theory we learned that for a given spin I, there are 2I + 1 quantized energy levels and there are 2I + 1 quantized states along B_0. Since we shall, for the most part, be using classical pictures, we would like to think that the classical and quantum mechanical pictures are both equally valid. It has been shown (12) that in spite of the quantum restrictions, the quantum mechanical expectation value (or quantum mechanical average) of μ, which is all that can be measured, behaves in its time dependence exactly as does **M** in the classical treatment that we are using. A second point concerns <u>saturation</u>, in which (for I=1/2) the populations of the upper and lower energy levels are equal. One might be tempted to equate a 90° pulse, which leaves no z' component of the magnetization, with saturation, which also leaves no net z' magnetization. The two processes are very different, however. In a cw saturation situation there is no net magnetization in any direction, and the recovery to an equilibrium population distribution occurs with a time constant T_1. With a 90° pulse (or with adiabatic rapid passage) the magnetization is moved from its equilibrium position along z' in a time so short that essentially no relaxation occurs, and a magnetization $\mathbf{M}_0$ is left along y'. Even though the quantized levels along $\mathbf{B}_0$ are no longer present there is a large component of phase-coherent magnetization in the rotating frame. In this case, a large component of magnetization can very quickly be created along $\mathbf{B}_0$ by a subsequent 270° pulse or by other pulse sequences (discussed in Chapter 2). These other pulses must, of course, be applied in a time less than T_2. In a true saturation situation no amount of pulse gymnastics can create a magnetization in a time substantially shorter than T_1. In cw experiments saturation has a reasonably well defined meaning, but it can be somewhat misleading in some pulse experiments. The approach which we shall take here is to treat $\mathbf{M}$ as a measurable and easily pictured quantity and to avoid, as far as possible, detailed discussions of quantized conditions.

1.5 Fourier Transform Spectroscopy

During the past decade transform techniques have been revolutionizing many forms of spectroscopy including NMR, infrared, and mass spectroscopy. The power of the new methods lies in the fact that they sample all resolution elements in a given spectrum simultaneously . In addition, the irradiation source can be more intense since no monochrometer or slits are required. In many ways Fourier transform (FT) NMR is very similar to FT-mass spectroscopy and FT-infrared spectroscopy. In FT-NMR the sample is irradiated with a short, intense, polychromatic burst, or pulse, of r.f. energy.

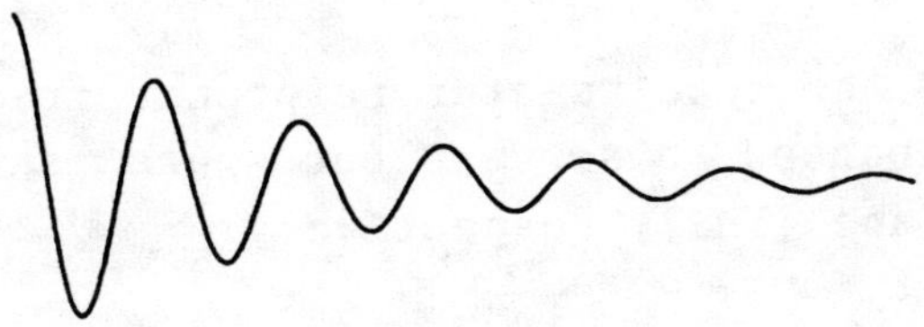

Figure 1.9 The FID signal for a sample of water

The result is that all nuclei in a given sample are caused to broadcast at their characteristic NMR frequencies simultaneously. If only a single frequency is present, then the time response signal following the r.f. pulse is a simple cosine wave which decays with a time constant, T_2 (see Figure 1.9). The frequency, v, of this signal can be determined very easily by measuring the period, t, of the sine wave, v = 1/t. Such a proton NMR signal arises for simple, pure samples such as water, benzene, or chloroform. In each of these cases the time response is a simple, damped cosine wave whose frequency is the difference between the proton

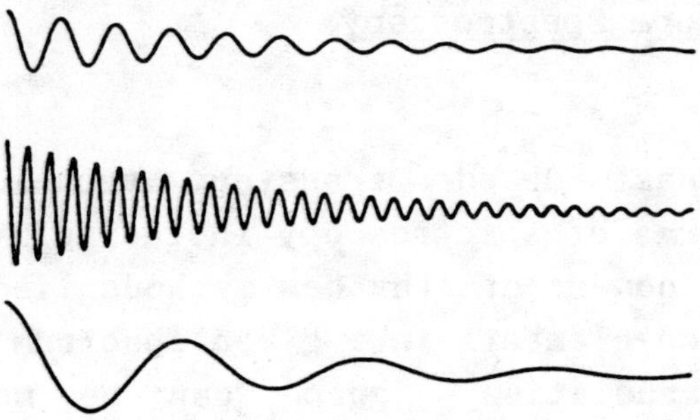

Figure 1.10 The FID signals for three separate samples. The top FID is for water, the middle FID for benzene and the bottom for TMS.

precessional frequency and the spectrometer reference frequency. Figure 1.10a-c shows the time responses observed for pure samples of water, benzene and tetramethylsilane (TMS), respectively.

In a more typical sample, for example, in a mixture of chloroform and benzene there are two frequencies present (see Figure 1.11a,b), and the time response that we see is simply the sum of these two signals (see Figure 1.11c). In Figure 1.12a-c we show the individual sine waves generated by three signals of relative intensities 1:2:1, which are typical of the triplet signals often observed in an NMR experiment. Figure 1.12d shows the sum of these three signals. This composite signal is quite typical of many of the FID's observed in an NMR spectrometer. A 1:3:3:1 quartet is quite similar. In all of the figures we have used an exponential damping factor of about 1.0 sec to approximate a typical NMR FID signal more closely . It is clear that even in these very simple cases it is rather difficult to look at the composite time response, or FID, signal to determine what signals at the various frequencies are present, or to determine the relative intensities of those signals. The Fourier transform is used to sort out the frequencies present and their relative intensities.

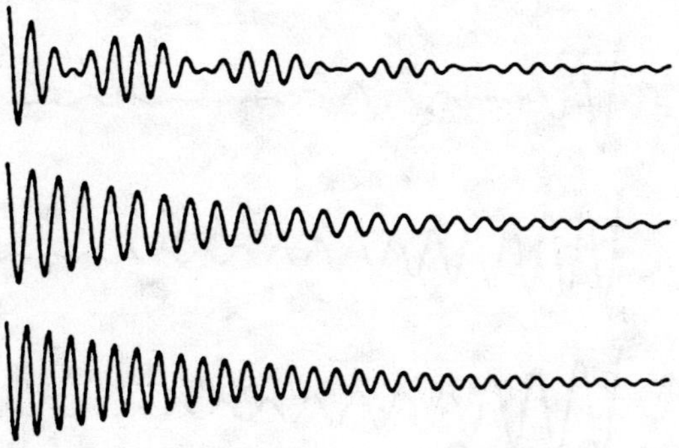

Figure 1.11 The middle FID is for a chloroform sample and the bottom FID is for a benzene sample. The top FID is for a sample of a mixture of 85% chloroform and 15% benzene.

The advantages of pulse, time response, or interferometric techniques in spectroscopy are that they allow one to obtain spectroscopic information very rapidly and with much higher sensitivity than the older cw, or dispersive, methods. But the information is presented to us in such a form that it is difficult to interpret. The Fourier transformation affords us a simple, straightforward and powerful method for converting the time response or interferogram to the more conventional and more easily interpretable frequency domain spectrum.

It is well known that a function f(t) can be expressed as a Fourier series, that is, as an infinite series of sines and cosines:

$$f(t) = \sum_n A_n \cos(n\pi/T)t + \sum_n B_n \sin(n\pi/T)t. \qquad (1.63)$$

It is shown in standard mathematical texts (14, 15) that this expansion is valid over the region $-T \leq t \leq T$, and formulas are derived for determining the coefficients A_n and B_n. If f(t) is a symmetric or <u>even</u> function [i.e., if $f(-t) = f(t)$], then all $B_n = 0$, and only the cosine series is needed. If f(t) is antisymmetric or <u>odd</u> [i.e., $f(-t) = -f(t)$], then only the sine series is needed. A general, asymmetric function requires the complete series. In all of our work it will be necessary to use the complete series.

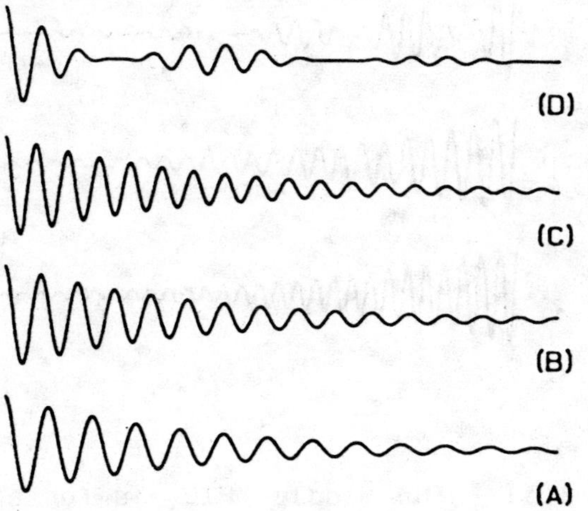

Figure 1.12 The bottom FID signals are for waveforms with frequencies (A) 1000 Hz, (B) 1200 Hz & (C) 1400 Hz. The top FID is the sum of the three bottom signals. D = A + 2B + C

If we mark off time in Figure 1.13a into N equal intervals and then measure the amplitude of the time response g(t) at each of these points, we can convert the time domain signal to the frequency domain signal $f(\omega)$ by the process:

$$f(\omega_j) = \Sigma_k \; g(t_k) \exp(-2\pi ijk/N), \quad j=0,1...N-1, \qquad (1.64)$$

where

$f(\omega_j)$ = amplitude of the jth point in the frequency domain
$g(t_k)$ = amplitude of the kth point in the time domain
N = the number of data points

Execution of this straightforward, but tedious, procedure for f(t) in Figure 1.13a results in the frequency domain spectrum shown in Figure 1.13b. This is the essence of the Fourier transform method. We will consider it in more detail later in Chapter 5.

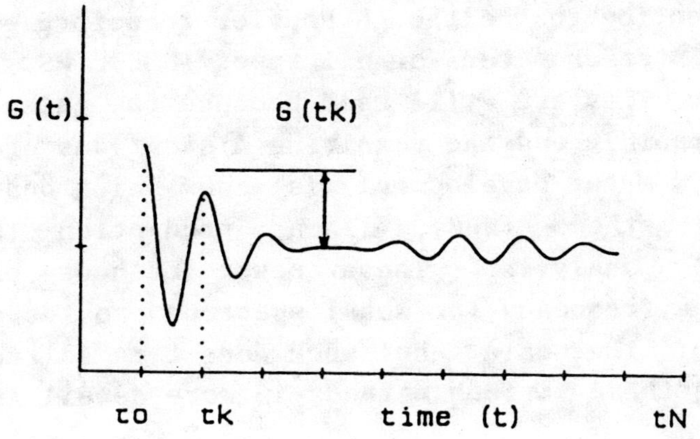

Figure 1.13a

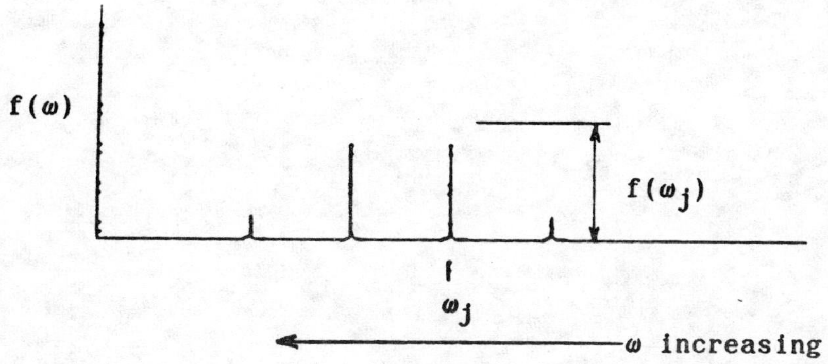

Figure 1.13b

Figure 1.13 (a) The pulsed spectrometer generates a time response G(t) as a function of time. After Fourier transformation of G(t) one obtains the frequency spectrum F(ω).

We should mention that although Fourier transform methods are widely used in current interferometers of all types (NMR, MS, and IR), they do distort the data. They are still used because the calculations involved can be perfomed rapidly and the resulting distortions are, in most cases, not very large. Much development is now being done in the areas of maximum entropy methods (MEM), linear prediction (LP), and auto-regressive (AR) analysis. These newer methods of converting an interferogram to a frequency (or mass) spectrum provide a more faithful representation of the data, but much more time is required to perform them. We discuss these various methods in more detail in Chapter 5.

CHAPTER TWO

FUNDAMENTAL NMR PULSE EXPERIMENTS

With most modern NMR spectrometers an almost bewildering variety of complex pulse experiments can be executed, even by the neophyte. These experiments can be most easily understood if one has a basic understanding of a few rather simple pulsed experiments. In this chapter we consider how simple spin systems respond to a variety of r.f. pulses of different phases and/or duration. We shall assume that the r.f. power from the r.f. power amplifier (or transmitter) can be switched on and off very rapidly and that the envelope of the pulse is a square wave. These conditions are easily met in almost all present-day (1986) commercial FT-NMR spectrometers. In addition, we assume that the pulse duration, t_p, is short relative to T_1 and T_2. That is, we assume that no relaxation occurs during the pulse time, t_p. Now recall that the angle through which the magnetization **M** moves during the pulse is given by

$$\theta = \gamma B_1 t_p \ . \qquad (2.1)$$

NOTE: If $t(\pi/2) = 1.0$ sec, then $\gamma B_1/2\pi = 250{,}000$ Hz.

In the same way in which we described the motion of **M** in Chapter 1, we can describe in a general way the interaction of **M** with its environment by the equation

$$d\mathbf{M}/dt = \gamma \, \mathbf{M} \times \mathbf{B} \ . \qquad (2.2)$$

The magnetic force, **B**, acting on **M** can arise from a number of sources:

1) from an electromagnetic field generated by an r.f. coil
2) from a large dc magnetic field
3) from fluctuating magnetic fields generated by unpaired electron motions
4) from fluctuating fields generated by the motions of magnetic nuclei

In a typical NMR experiment, several of these forces may be acting on the **M** vector at the same time. Let us assume that we are already at equilibrium and that **M** is in a stationary state in which it is along the z' axis in the rotating reference frame. If we apply a r.f. field $\mathbf{B}_1$ along the x' axis, **M** will precess about $\mathbf{B}_1$ in the same way that a gyroscopic top precesses in the earth's gravitational field (remember that the nuclei possess both a magnetic moment <u>and</u> angular momentum); see Figure 2.1.

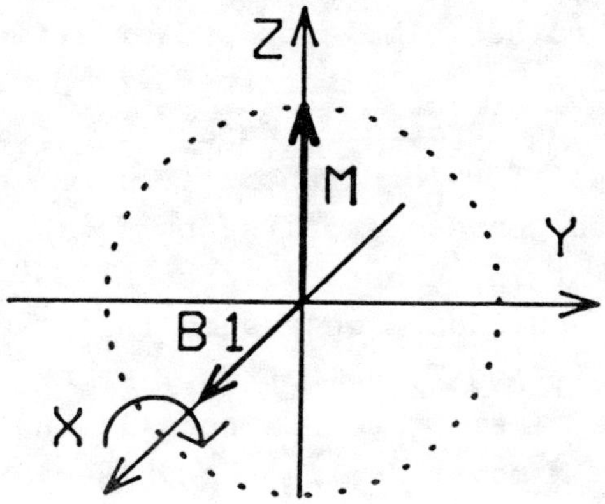

Figure 2.1 If a γB_1 pulse is applied along the x axis in the rotating frame, the macroscopic magnetization, **M**, will rotate about the x axis at a rate given by $d\mathbf{M}/dt = \gamma\, \mathbf{M} \times \mathbf{B}_1$.

We can always keep track of the direction in which **M** moves by remembering the right-hand rule: if the fingers of the right hand point from the direction of **M** to the direction along which $\mathbf{B}_1$ is directed, then the thumb will point in the direction in which **M** will move. Thus if $\mathbf{B}_1$ is directed along the +x' axis, **M** will rotate clockwise about the +x' axis (see Figure 2.2a). If $\mathbf{B}_1$ is directed along the -x' axis, then, of course, **M** will rotate in the opposite direction. This is a <u>very</u> important concept and has far-reaching consequences, for it affects the NMR signals which we obtain from our spectrometer. And it provides the basis for the four basic spin-gymnastic exercises which are shown schematically in Figure 2.2a-d.

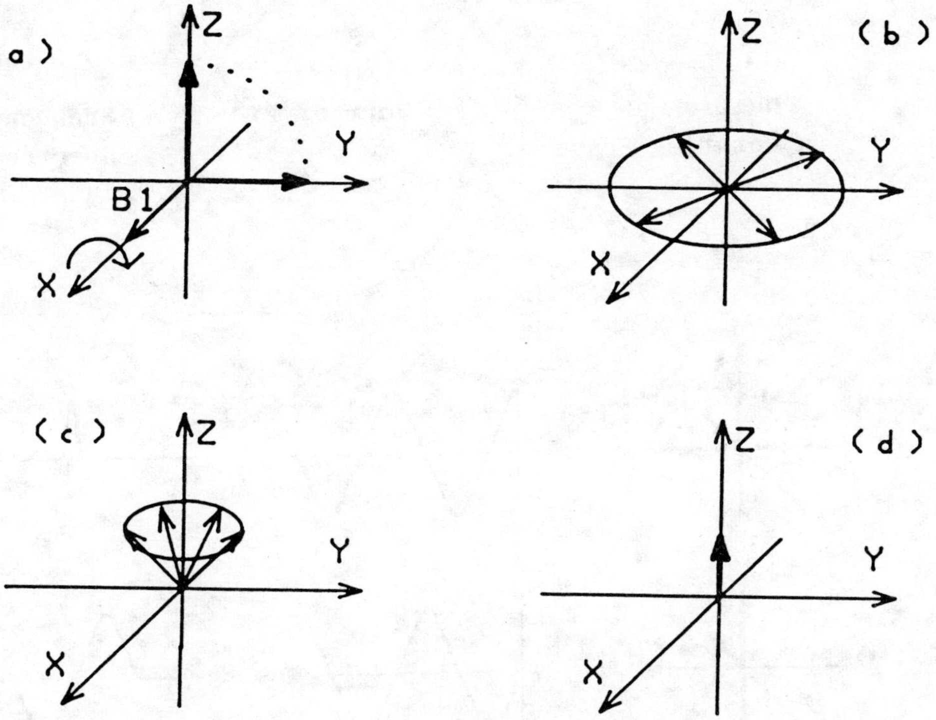

Figure 2.2 (a) A 90^{0} γB_1 pulse along x rotates **M** to a position along the y axis in the rotating frame. (b) The magnetization, **M**, dephases in the xy plane due to magnetic field inhomogeneities and T_2 relaxation. (c) A longitudinal magnetization will be re-established due to T_1 relaxation. (d) At a time, $\tau = 5T_1$, the full macroscopic magnetization is again along the z axis.

We shall hereafter use the notation for pulses given in Table 2.1 The letter +/- x or +/- y denotes the axis along which the pulse of r.f. energy is applied, and is preceded by the pulse angle in units of radians. Thus $\pi/2(+x)$ denotes a $\pi/2$ (or 90^{0}) pulse along the positive x axis in the rotating frame. By a careful choice of the B_1 axis and the duration of the B_1 pulse we can put **M** at any arbitrary position in our rotating frame of reference. The position of **M** in the xy-plane determines the nature of the signal which we observe in the NMR spectrometer. If **M** lies along the +y axis, we obtain a positive FID signal as shown in Table 2.1a. If **M** lies along the -y axis, we obtain a negative FID signal, and so on (see Table 2.1).

Table 2.1

γB_1 Axis	Position of **M** in Rotating Frame	Form of FID	Frequency Spectrum	
+x	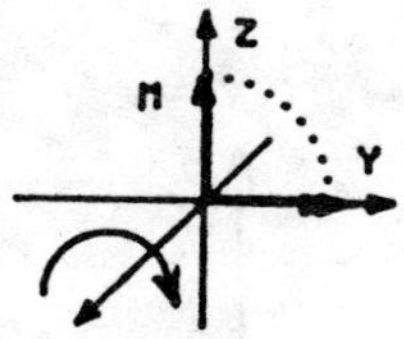	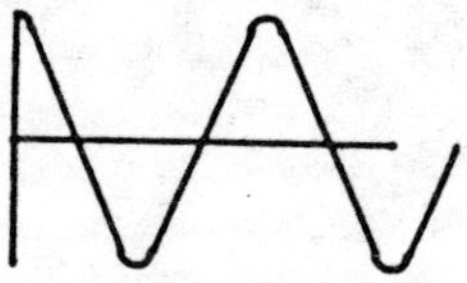	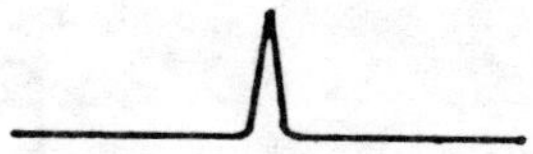	(a)
+y	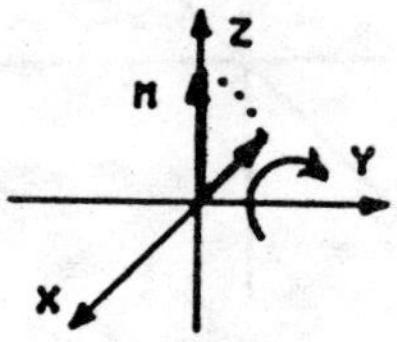	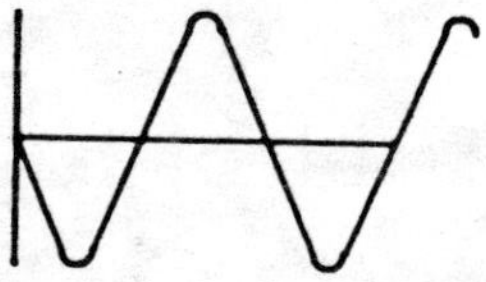	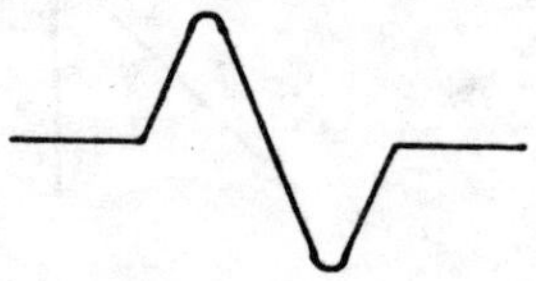	(b)
-x	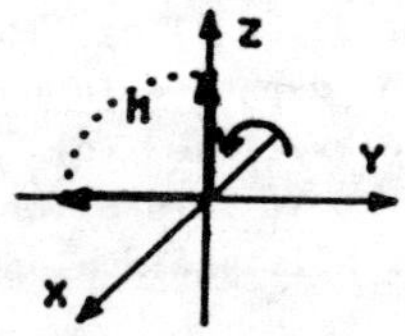	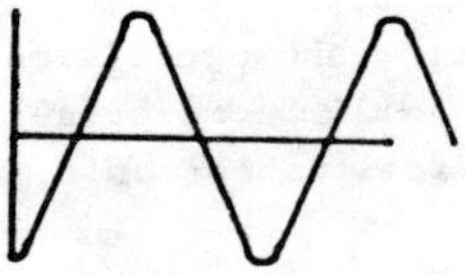	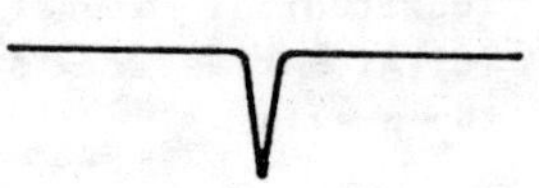	(c)
-y	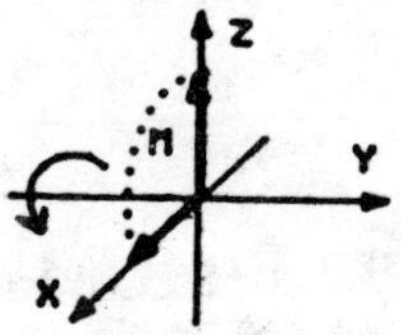	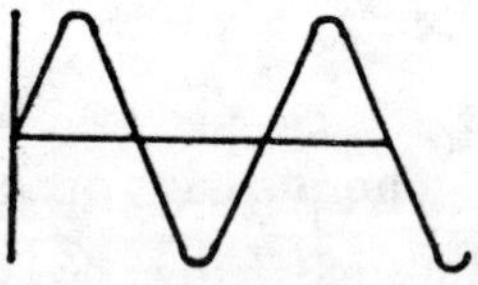	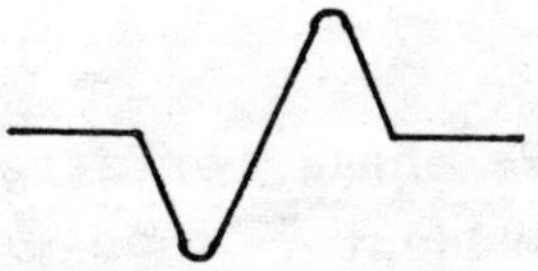	(d)

Table 2.1 (a) If the magnetization, **M**, is rotated to a position along the y axis in the rotating frame by a 90° γB_1 pulse along the x axis, the FID signal will be a cosine wave. The Fourier transform of this FID will be an absorption spectrum. (b) If γB_1 is applied along the +y axis, **M** will be rotated to a position along the -x axis. The FID signal will be 90° out of phase with the FID signal of (a) and the FFT of this FID will be a dispersion spectrum. (c) If γB_1 is along -x, the FID will be 180° out of phase with the FID in (a) and the FFT will be a negative absorption signal. (d) Similarly, a γB_1 pulse along -y will give a cosine wave 270° out of phase with the FID in (a). and the FFT of this signal will result in a negative dispersion signal in the frequency domain.

We have already seen that such nuclear spin gymnastics are quantum processes that take place essentially instantaneously, provided that γB_1 is large enough. The position of **M** at any time can be quite accurately predicted by using equations 2.1 and 2.2. Let us now expand equation 2.2 in order to see more clearly how the individual vector components are affected:

$$
\begin{aligned}
d\mathbf{M}/dt &= \gamma\, \mathbf{M} \times \mathbf{B} \\
&= \gamma[\mathbf{i}(M_yB_z - M_zB_y) + \mathbf{j}(M_zB_x - M_xB_z) \qquad (2.3) \\
&\quad + \mathbf{k}(M_xB_y - M_yB_x)] \; .
\end{aligned}
$$

And in terms of the individual vector components,

$$dM_x/dt = \gamma\,(M_yB_z - M_zB_y), \qquad (2.4a)$$

$$dM_y/dt = \gamma\,(M_zB_x - M_xB_z), \qquad (2.4b)$$

$$dM_z/dt = \gamma\,(M_xB_y - M_yB_x), \qquad (2.4c)$$

But in the rotating frame B_o is compensated by a fictitious field, and the resulting field is zero. $\gamma B_1 = \gamma B_{1x}$, so in the rotating frame

$$dM_x/dt = 0, \quad dM_y/dt = \gamma B_{1x}, \quad dM_z = -\gamma M_y \gamma B_{1x}. \qquad (2.5)$$

This is a fairly common situation, often encountered in elementary physics. The equations of motion of M are described by two coupled, linear, first-order differential equations. It is left as an exercise to the student to show that the solution to this set of equations is

$$M_y = \gamma \sin(kt), \qquad (2.6a)$$

$$M_z = -\gamma \cos(kt), \qquad (2.6b)$$

These equations express mathematically what we have described earlier with simple physical pictures. Under the influence of a γB_{1x} field in the rotating frame of reference, there is no component of **M** along the M_x axis and none develops. The components of M_y and M_z vary in a sinusoidal fashion, and are 90° out of phase with one another, as shown in Figure 2.3.

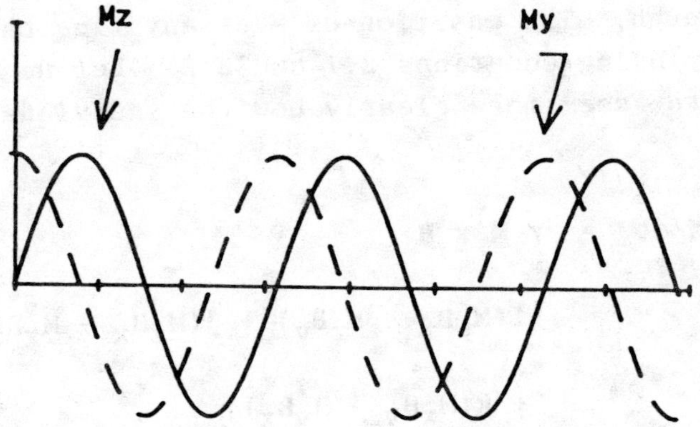

Figure 2.3 Under the influence of a γB_1 field along the x axis in the rotating frame, the magnetization **M** will rotate about the x axis. The components M_y and M_z will oscillate as sine waves 90° out of phase with one another.

Here is clear mathematical proof, if any is needed, that **M** rotates at a constant rate in the zy plane under the influence of a r.f. field, γB_{1x}. Similar calculations confirm the other sketches which we have given in Figure 2.2. Thus, equation 2.3 describes how **M** moves under the influence of any magnetic field. We can also use this simple equation to gain some insight into how molecular motions are related to T_1 and T_2 relaxation processes.

2.1 Spin-Lattice (T_1) and Spin-Spin (T_2) Relaxation Times

The primary relaxation mechanism for most spin-1/2 nuclei in molecules in liquids is the magnetic dipole-dipole interaction. As a typical example, consider two protons, A and B. The total field at A arises from the large magnetic field, B_o, and from the field, $\mathbf{B}_{loc}$, arising from nucleus B

$$| \mathbf{B}_{loc} | = \mu_B (3 \cos^2 \theta - 1)/r_{AB}^3 . \qquad (2.7)$$

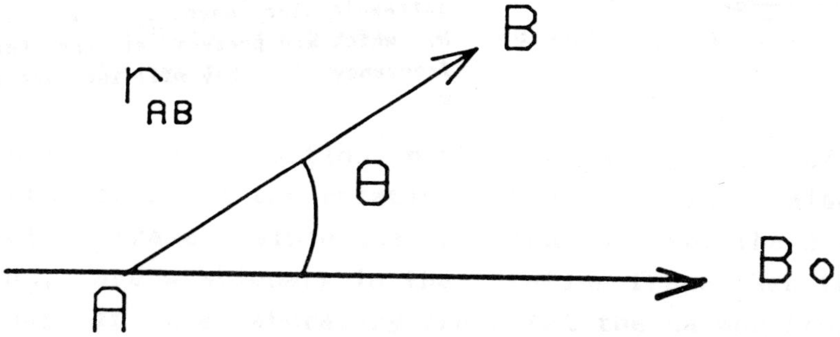

Figure 2.4 The local magnetic field at nucleus A, B_{loc}, arising from a neighboring nucleus B which is located at a distance r_{AB} is given by $B_{loc} = \mu_B (3 \cos^2 \theta - 1)/r^3$, where μ_B is the magnetic moment of nucleus B and θ is the angle which the A-B internuclear vector makes with the dc magnetic field, B_o.

If the protons A and B are in the same molecule (e.g., in a water molecule) and if the molecule rotates, the angle θ will change. This causes the local field , $\mathbf{B}_{loc}$, to fluctuate (the field fluctuation can be anywhere from milligauss to about 160 gauss). And for such rotational motions, when we speak of a molecular correlation time, τ_c, we mean the time required for the angle θ (see Figure 2.4) to change appreciably. It is this coupling between the nuclear precession and the lattice (or molecular) motions mediated by the nuclear dipole-dipole interaction that allows thermodynamic heat transfer to take place between the nuclear spin system and the rotational motion of the molecules. But the relaxation takes place only if the field, $\mathbf{B}_{loc}$, at nucleus A changes, and only if the changes take place with frequency components at the Larmor frequency. The origin of these magnetic fluctuations is not important; nucleus A neither knows nor cares about the source of the fluctuations.

T_1, since they are affected by both fast and slow dynamic processes. On the other hand, T_1 is affected only by rapid dynamic processes, those that occur at the rate of $1/\omega_0$ or faster. We shall encounter many examples of these differences. The important points to remember are:

a) Fluctuations of the magnetic field at one nucleus which are generated by the molecular motions of the molecule or the molecular motions of a neighboring magnetic nucleus cause relaxation (that is, a transfer of energy from the spin system to the lattice).

b) Both fast and slow fluctuations contribute to spin-spin relaxation (T_2), but only fast fluctuations contribute to spin-lattice relaxation (T_1).

c) Consequently, T_2 is always equal to or less than T_1.

We shall present a more detailed explanation of T_1 and T_2, and will diccuss relaxation mechanisms, in Chapter 4. At this point we discuss how T_1, T_2 and the closely related $T_{1\rho}$ (the spin-lattice relaxation time in the rotating frame) can be measured experimentally.

2.2 Free Induction Decay and the Measurement of the Spin Lattice Relaxation Time, T_1

In cw NMR spectroscopy, the spin-lattice relaxation time, T_1 is not of prime importance. In pulse NMR spectroscopy, however, T_1 and T_2 are of paramount importance. With cw methods the measurement of T_1 is difficult and not very accurate, especially for relaxation times less than a few seconds. Pulse methods, on the other hand, afford one a straightforward, relatively easy, and accurate way to measure T_1. The most common experiment used to measure T_1 is the inversion recovery or 180°-τ-90° sequence. This experiment is depicted in Figure 2.6. At time t=0, the magnetization vector, **M**, is inverted by a 180° pulse. After this 180° pulse, **M** lies along the negative z' axis. Because of spin-lattice relaxation (this is sometimes called longitudinal relaxation), **M** will increase in value from $-M_o$ through zero and back to its full equilibrium value of $+ M_o$. If at some time, τ, after we apply the 180° pulse, we apply a 90° pulse, **M** is rotated by an additional 90° about the x' axis. **M** will then lie somewhere along the y' axis. If τ is short compared to T_1, **M** will lie along the negative y' axis and a negative free induction decay (FID) signal will be seen. If τ is long compared to T_1, then M will lie along the positive y' axis and a positive FID will be seen. The initial height of the FID signal (or the intensity of the line following the FFT of the FID signal) is proportional to the magnitude of **M**, and hence to M_z at time τ. If we wait for a period equal to at least five times T_1 and repeat this experiment for different τ times, we can monitor how rapidly **M** recovers to its equilibrium position along the z axis. If we plot the magnitude of **M** versus the time, τ, between the pulses, we obtain the curve shown in Figure 2.6c. Quantitatively, the recovery of M_z is given by

$$dM_z/dt = -(M_z - M_o)/T_1. \qquad (2.10)$$

Straightforward integration of this equation (with $M_z = -M_o$ at $t = 0$) yields the result

$$M_z = M_o[1 - 2 \exp(-t/T_1)]. \qquad (2.11)$$

This is the function that is shown in Figure 2.6c. In actual practice it is more useful to plot the data in the form

$$\ln (A_\infty - A_\tau) = \ln(2A_\infty) - \tau/T_1. \qquad (2.12)$$

where A_τ is the intensity of the signal following the 90^0 pulse at time τ and A_∞ is the limiting value of A_τ for a very long interval between the 180^0 and 90^0 pulses. T_1 is determined from the slope of the plot of $\ln(A_\infty - A_\tau)$ versus τ. More details of this experiment are discussed in Chapter 4.

It can be seen from equation 2.12 that when $\tau = T_1\ln2$ seconds = $0.692T_1$ seconds there will be no signal, that is, A_τ is zero for τ = $0.692T_1$ seconds. This time is sometimes called the "null time." It is very tempting to try to measure T_1 by finding this null time and multiplying by 1.44 ($T_1 = 1.44\ \tau_{null} = \tau_{null}/\ln2$). This process will provide a rough estimate of the value of T_1, but it can lead to very large, serious errors. We shall discuss this point in more detail in Chapter 3 on instrumentation requirements.

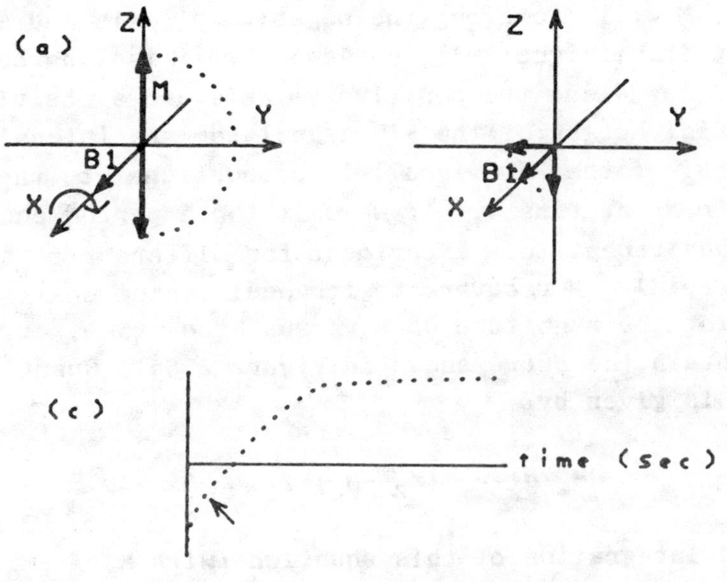

Figure 2.6 One of the methods for determining the value for T_1 is the 180^0-τ-90^0 sequence. (a) First, M is inverted by a π pulse at time t=0. (b) After a delay of τ seconds a $\pi/2$ sampling pulse is used to measure the recovery of M. If τ is short compared to T_1, the $\pi/2$ pulse will rotate M to a position along the -y axis and the FID signal (and the absorption signal after Fourier transformation) will be negative. Such a condition is shown by the arrow in (c).

2.3 Refocusing Experiments; T_2 and $T_{1\rho}$ Experiments

Since $T_2 \leq T_1$, one might think that the measurement of the width of a line might be a good measure of T_2 (see equation 1.11). However, this is true only when $T_2 \ll T_2^*$ (where T_2^* is related to the magnetic field homogeneity; see equation 1.11). A much better way to measure T_2 was first proposed by Hahn (5), who named the experiment a "spin-echo" experiment. This experiment consists of two pulses, a $\pi/2(+x)-\tau-\pi(+x)$. At a time 2τ an "echo" is seen. The rationale of this method is shown in Figure 2.7, which shows the behavior of the magnetization in the rotating frame. In (a) **M** is shown being rotated through 90° by the application of $\mathbf{B}_1$ along the positive x' axis. The total magnetization **M** can be thought of as the vector sum of individual macroscopic magnetizations $\mathbf{m}_i$ arising from nuclei in different parts of the sample and hence experiencing slightly different values of the applied field, which is never perfectly homogeneous. There is thus a range of precession frequencies centered about v_o, which we have taken as the rotation frequency of the rotating frame. In (b), then, the $\mathbf{m}_i$ begin to fan out, as some nuclei precess faster and some slower than the frame. At a time τ after the 90° pulse, a 180° pulse is applied, also along the positive x' axis, as shown in (c). The effect of this pulse is to rotate each $\mathbf{m}_i$ by 180° about the x' axis. Thus those $\mathbf{m}_i$ that are moving faster than the frame (shown in [b] moving toward the observer or clockwise looking down the z' axis) naturally continue to move faster, but in (d) their motion is now away from the observer. At time 2τ all $\mathbf{m}_i$ come into phase along the negative y' axis, as shown in (e). The continuing movement of the $\mathbf{m}_i$ causes them again to lose phase coherence in (f).

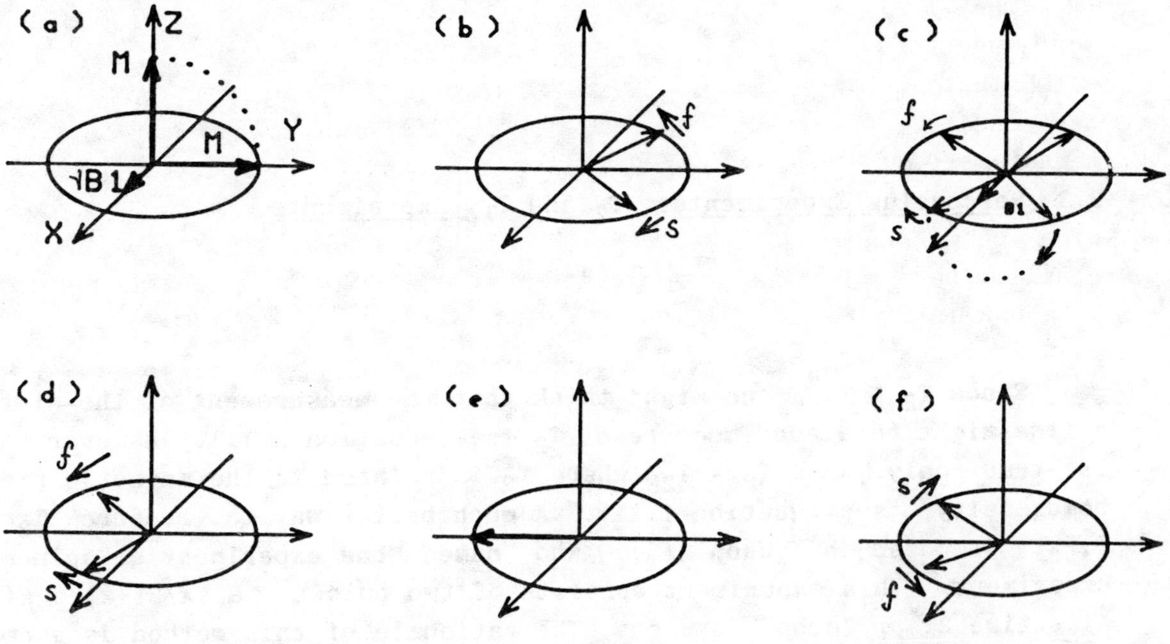

Figure 2.7 The Hahn spin-echo experiment. (a) A $\pi/2$ pulse along the x axis rotates **M** to a position along the y axis in the rotating frame. (b) **M** dephases as a result of magnetic field inhomogeneities and T_2 precesses. Those nuclei precessing faster than average (because of field inhomogeneities) appear in the rotating frame to move toward the observer. Looking down on the positive z axis, they appear to move clockwise. Those moving more slowly than average move in a counterclockwise direction. (c) A π pulse along the x axis at a time τ after the initial $\pi/2$ pulse rotates the magnetization about the x axis and initiates the refocusing of the individual microscopic magnetizations. (d) If there is no molecular diffusion, the faster nuclei will still be moving in a clockwise direction and now move away from the observer. The slower nuclei, similarly, move counterclockwise toward the observer. (e) At a time 2τ after the initial $\pi/2$ pulse the magnetization is refocused along the negative y axis. (f) At a time greater than 2τ the dephasing continues.

The rephasing of the $\mathbf{m}_i$ causes a free induction signal to build to a maximum at 2τ, but the signal is, of course, negative relative to the initial free induction decay, since rephasing of the $\mathbf{m}_i$ occurs along the <u>negative</u> y' axis. If the field were perfectly homogeneous, the echo amplitude might be just as large as the initial value of the free induction following the 90^0 pulse. However, the $\mathbf{m}_i$ vectors lose phase coherence not only because of field inhomogeneity, but also because of the natural processes responsible for transverse relaxation. These "natural" processes, however, are irreversible and cannot be refocused.

Consequently, the echo amplitude will decay in time with a time constant, T_2. That is, T_2, may in principle be determined from a plot of peak echo amplitude versus the time, τ, between the $\pi/2$ and π pulses. It should be noted that the amplitude of the echo is not affected by magnetic field homogeneity, provided that diffusion is relatively slow. As in the measurement of T_1 by the 180°-τ-90° sequence, it is necessary to carry out a separate pulse sequence for each value of τ and to wait an adequate time between pulse sequences (at least five times T_1) for restoration of equilibrium. A typical set of data for this type of spin-echo experiment is shown in Figure 2.8.

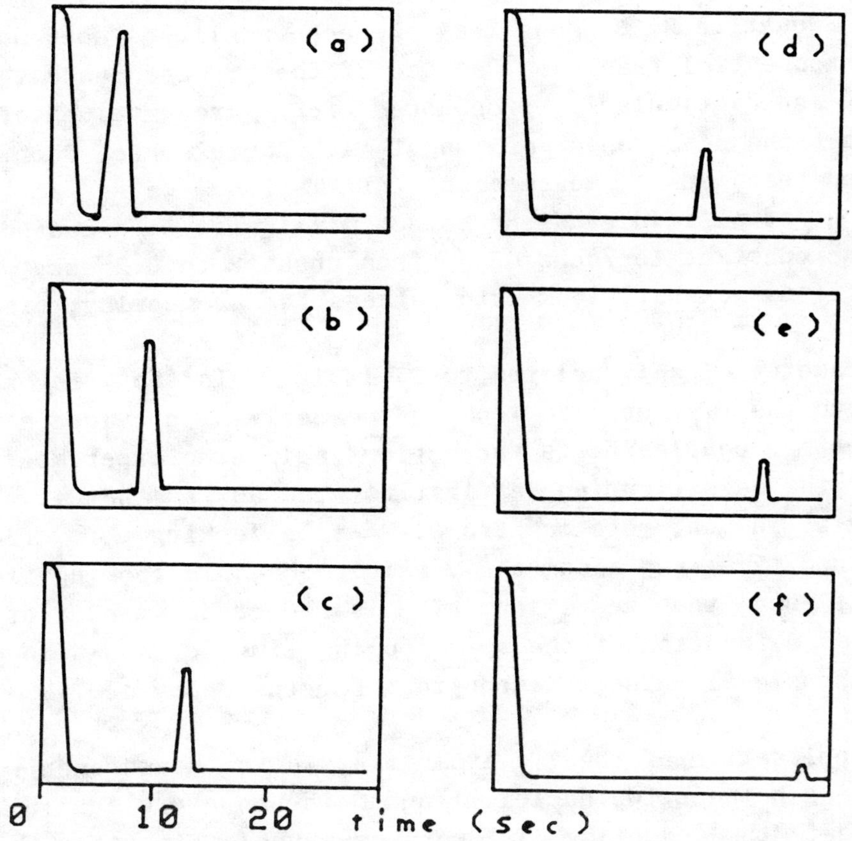

Figure 2.8 In a typical spin-echo experiment the time τ between the initial $\pi/2$ pulse and the refocusing π pulse is varied from about 0.1 second to about 1.0 second. The amplitude of the echo decreases and the time τ between the pulses increases. In this experiment the echoes are all positive since a diode detector was used in the receiver.

The spin-echo technique, as we have described it, is limited in its range of applicability because of the effect of molecular diffusion. The precise refocusing of all the m_i is dependent upon each nucleus remaining in a constant magnetic field during the time of the experiment (2τ). If diffusion causes nuclei to move from one part of an inhomogeneous field to another, the echo amplitude is reduced. The effect of diffusion in the spin-echo experiment is dependent upon the spatial magnetic field gradients (G), the diffusion coefficient (D), and the time during which diffusion can occur. It has been shown (16) that the amplitude of the echo for a pulse separation τ is

$$E(2\tau) = k_e \exp[-(2\tau/T_2) - (2/3)\gamma^2 G^2 D \tau^3], \qquad (2.13)$$

where $E(2\tau)$ is the echo amplitude at time = 2τ and k_e is a normalization constant. Equation 2.13 shows that the echo amplitude does not decay in a simple exponential fashion. Because of the τ^3 dependence the effect of diffusion is particularly pronounced for large values of τ and thus affects most the measurement of long T_2's. For example, Carr and Purcell (17) showed that in a measurement of T_2 for water at 25° C the application of a field gradient of 280 milligauss cm^{-1} caused a decrease in the time constant for echo decay from about 2 to 0.2 sec. In the following sections we shall look into methods for overcoming this problem.

A parameter closely related to T_2 is $T_{1\rho}$. In fact, for most liquids $T_{1\rho} = T_2$ and the $T_{1\rho}$, or spin-locking, experiment provides a very useful alternative or complement to the various spin echo experiments. In this experiment the magnetization is first rotated 90° so that it lies along the y axis in the rotating frame. A spin-locking r.f. field is then applied for a time, τ, along the y axis. The spin-locking field is then turned off and the resulting FID is recorded. A plot of the signal amplitude as a function of the spin-locking time, τ, shows an exponential decay of the signal. The time constant for the decay is $T_{1\rho}$.

One application of the $T_{1\rho}$ experiment is the measurement of scalar spin coupling between two nuclei, I and S when one nucleus (S) relaxes so rapidly that the coupling is not observable in the ordinary high resolution experiment. If T_1 for S can be measured, equations 4.29 and 4.30 can be used to obtain the spin coupling constant J_{IS}. In some cases T_1 is too short to be measured directly. In such a case the $T_{1\rho}$ experiment can still provide the necessary information. In the rotating frame it is straight forward to show that

$$R_{1\rho}(I) = A^2 \frac{S(S + 1)}{3} \frac{T_1(S)}{1 + \omega_1^2 \; [T_1(S)]^2} \qquad (2.14)$$

Here $\omega_1 = \gamma_I B_1$, and the equation holds for $B_1 << B_o$ and $(\omega_I - \omega_S)T^2(S) >> 1$, two conditions that are normally met. By rearranging equation 2.14 into the form

$$T_{1\rho}(I) = \frac{3}{A^2 S(S+1) T_1(S)} + \frac{3T_1(S)}{A^2 S) S+1)} \omega_1^2 \qquad (2.15)$$

we see that a plot of $T_{1\rho}(I)$ vs. ω_1^2 gives a straight line. From the slope m and the intercept b of the line, we can obtain A and $T_1(S)$:

$$A^2 = \frac{3}{S(S + 1)(bm)^{0.5}} \qquad (2.16)$$

$$T_1(S) = (m/b)^{0.5} \qquad (2.17)$$

For a sample of $HMn(CO)_5$ in toluene at room temperature (50b), for example, the results are $J = A/2\pi = 15.9$ Hz and $T_1(Mn) = 86$ μsec.

2.4 The Carr-Purcell Experiment

Carr and Purcell (17) showed that a simple modification of Hahn's spin-echo method reduces drastically the effect of diffusion on the determination of T_2. This method may be described as a $\pi/2(+x)$, τ, $\pi(+x)$, 2τ, $\pi(+x)$, 2τ, $\pi(+x)$, 2τ, ..., sequence, or, more commonly, a **Carr-Purcell** sequence. As in the spin-echo method described previously, all pulses are applied along the positive x' axis. The result is the same as that shown in Figure 2.8, except that following the dephasing of the $\mathbf{m}_i$ shown in Figure 2.10f another $\pi(+x)$ pulse at time 3τ after the initial $\pi/2(+x)$ pulse causes a rephasing of the $\mathbf{m}_i$ along the positive y' axis at time 4τ. Subsequent $\pi(+x)$ pulses at 5τ, 7τ, 9τ, etc., cause echoes at 6τ, 8τ, 10τ, etc., alternately positive and negative, as shown in Figure 2.9a.

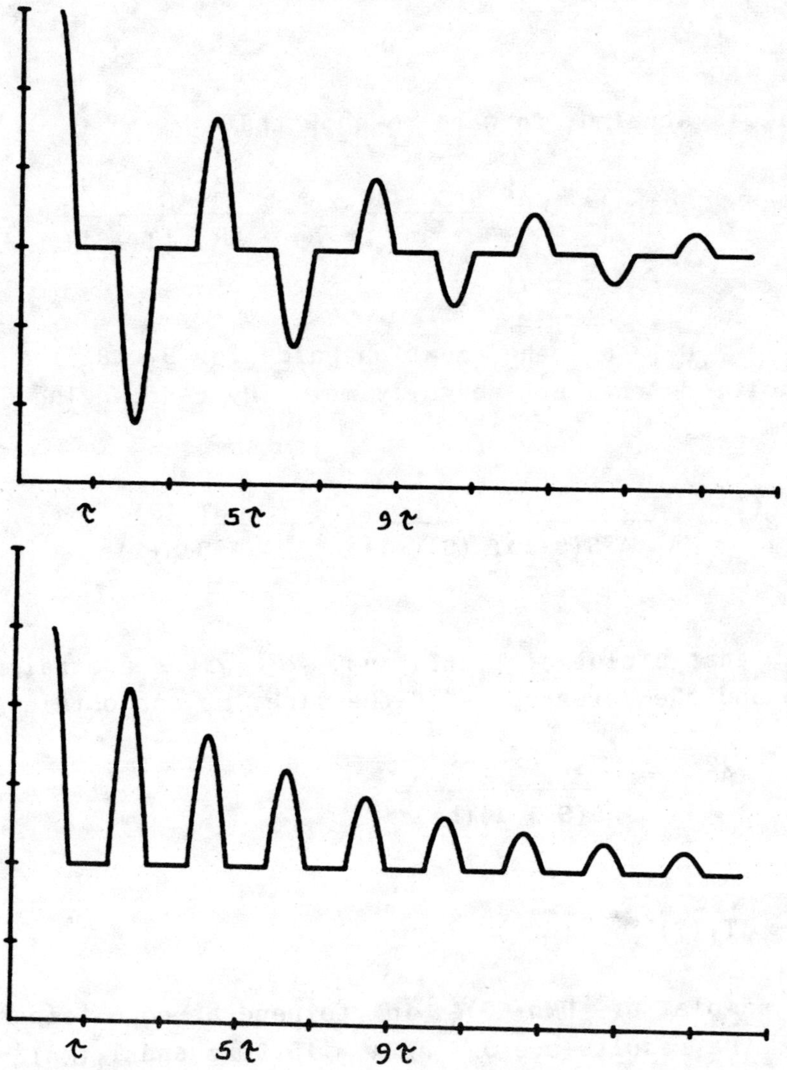

Figure 2.9 (a) In the Carr-Purcell experiment a series of π refocusing pulses are used at times τ, 3τ, 5τ, 7τ, and so on. This results in a series of echoes with alternating amplitude (and maximum intensities at 2τ, 4τ, 6τ, and so on) as the magnetization is refocused in an alternating fashion along the -y axis and the +y axis. (b) In the phase-shifted Carr-Purcell experiment the π refocusing pulses are along the +y axis and the refocusing is about the +y axis. The echoes decay more slowly because the y axis refocusing results in substantial cancellation of errors in setting the pulse widths. Since the τ interval can be made quite short, the effects of spin diffusion can be eliminated.

There are two distinct advantages of the Carr-Purcell method. First there is a considerable saving in time, for a train of n echoes may be obtained in a single sequence, whereas in the unmodified spin-echo method n sequences are required, with a waiting time between sequences that is long relative to T_1 (usually at least five times T_1 is the time between sequences). Second, the effect of diffusion may be virtually eliminated by making τ short, since it is only during a period 2τ that diffusion is effective in reducing the amplitude of an echo. For the Carr-Purcell sequence, equation 2.13 becomes

$$E(t) = k_e \exp[-(t/T_2) - (1/3)\gamma^2 G^2 D\tau^2 t] . \qquad (2.18)$$

As can be seen, by choosing τ short enough, the diffusion term may, in principle, be made as small as desired. The limitations are usually instrumental in nature; they will be discussed in more detail in Chapter 3.

Although the Carr-Purcell method is a great improvement over the simple Hahn spin-echo experiment, it is still subject to serious errors. As we have seen from equation 2.1, the angle through which **M** is rotated by a pulse depends upon B_1 and the pulse time t_p. The Carr-Purcell experiment, like most of the multipulse experiments we will discuss later in this book, requires that the pulses be reasonably accurate and that the the r.f pulses be uniform over the entire sample volume. In practice it is quite difficult to meet these experimental conditions, so we must find ways to compensate for them. One procedure for obtaining $\pi/2$ and π pulses is to adjust t_p experimentally in order to obtain a maximum FID (free induction decay) signal for the $\pi/2$ pulse and a minimum (ideally zero) FID signal for the π pulse. Actually, the $\pi/2$ pulse is not very sensitive to the duration of the pulse, t_p. Therefore, in practice one usually adjusts the π pulse for a minimum (the minimum is rather sharp) and sets the $\pi/2$ pulse width for half of the π pulse width. With this procedure the pulse widths can be rather accurately set, but for only a small part of the sample. Inhomogeneity over the sample makes it quite impossible in most cases to have accurately set pulses for the entire sample. The result is that the errors in setting the π pulses in a Carr-Purcell experiment accumulate and become quite troublesome. A common method to overcome this problem is to shift the phase of the π pulses. This experiment is sometimes called the PSCP (phase shifted Carr-Purcell) experiment or the Carr-Purcell-Meiboom-Gill (CPMG) experiment. This technique uses the same pulse sequence as the Carr-Purcell technique, but the π pulses are applied along the positive y' axis. Figure 2.10 shows the result for π pulses that are exactly $\pi(+y)$ radians. The $\mathbf{m}_i$ are now rotated about the y' axis, and all echoes refocus along the positive y' axis; hence only positive echo signals are obtained.

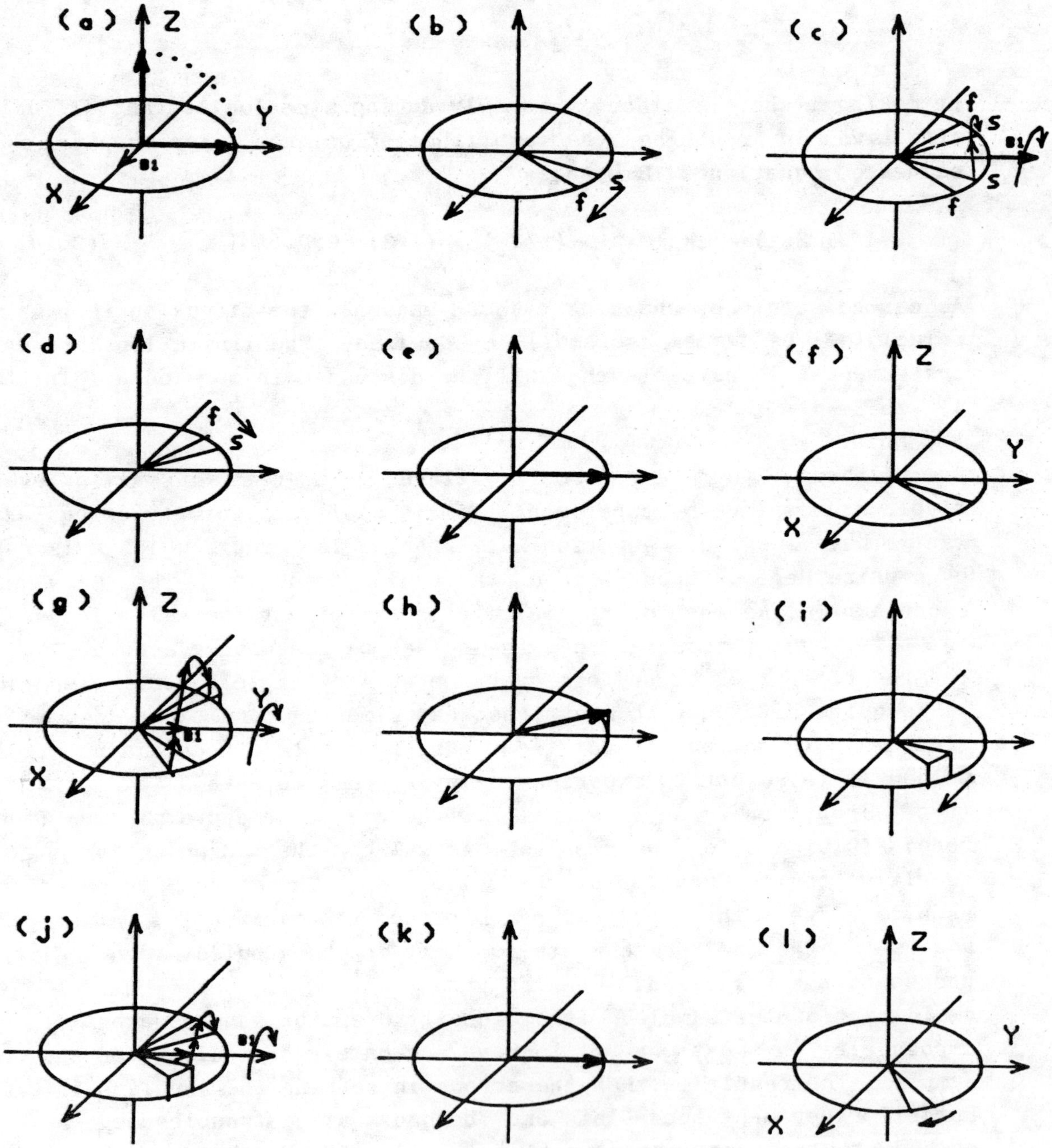

2.10 The phase-shifted Carr-Purcell experiment. In this experiment the refocusing takes place about the +y axis. In this example all nuclei are precessing at a rate a little faster than the rotating frame frequency, ω_0. (a) A $\pi/2$ pulse along the x axis rotates **M** along the +y axis. (b) Due to T_2 effects and magnetic field inhomogeneities the magnetization begins to dephase. (c) At a time τ after the initial $\pi/2$ pulse a π refocusing pulse is applied along the +y axis. (d) The nuclei begin to refocus (or rephase). (e) At a time 2τ the magnetization is refocused along the +y axis, but (f) continues to dephase after the refocusing. (g) For a pulse smaller than 180° (e.g., 170°), the magnetization will not be rotated down into the xy plane in the rotating frame. (h) The refocusing takes place above the xy plane (in this case 10° above the plane) and (i) the refocusing continues. (j) When the next 170° refocusing pulse is applied the magnetization is refocusing down into the xy plane (this time it started out 10° above the plane), and (k) at time 4τ the magnetization is refocused in the xy plane along the +y axis. (l) At time longer than 4τ the dephasing and refocusing continue with pulse errors being compensated at every second refocusing pulse.

If the pulse is slightly less that π radians (for example, $\pi(+y) - \delta$), the m_i rotate to a region above the x'y' plane. They then refocus above the plane, and consequently the projection along the y' axis (which is all that we can observe experimentally) is reduced. This in turn causes a decrease in the amplitude of the first echo which we observe. Now the m_i continue to dephase above the x'y' plane and the second $\pi(+y)-\delta$ pulse is now just long enough to cause them to rotate precisely into the x'y' plane. Thus the second and all even-numbered echoes have the correct amplitude, while the odd-numbered echoes are reduced slightly, but not cumulatively. Figure 2.11 shows a typical result.

While the CPMG experiment eliminates most of the problems of setting the pulse widths and problems with diffusion, other parameters must be carefully controlled if one expects to obtain meaningful values for T_2. For a complete and clear understanding of the T_2 experiment the reader is referred to an excellent paper by Vold, Vold, and Simon (18).

CHAPTER THREE

INSTRUMENTATION

The variety of experiments which can be executed by a modern NMR spectrometer is almost unlimited. In order to do these experiments properly a great deal of sophistication is required on the part of the FT-NMR instrument and on the part of the person using the instrument. The experiments range from the simplest one, obtaining a proton NMR spectrum, to very complex experiments requiring feats of dazzling spin gymnastics on the part of two or more different spins.

In this chapter we discuss a number of the basic instrumentation requirements for a modern pulse NMR spectrometer as well as a few examples of more sophisticated experiments. We also call attention to some of the more common limitations of the spectrometers and what, if anything, can be done to overcome the limitations.

3.1 The Basic Pulse NMR Spectrometer

The basic spectrometer consists of one or more r.f. transmitters for manipulating the various nuclear spins, an r.f. receiver for detecting the signal generated by the precessional motion of the spin magnetizations and a special transmitter-receiver unit to provide a field-frequency stabilization signal. For example, in a typical modern spectrometer one might "excite" (at 50 MHz) the carbon-13 nuclei in a sample by rotating the macroscopic magnetization to a direction perpendicular to the B_0 magnetic field and, then observe the signal induced in the receiver coil by the rotating, transverse carbon magnetization and, at the same time, decouple with a second r.f. transmitter (at 200 MHz) the proton spins. A third r.f. transmitter-receiver system operating at 30 MHz is used to excite and detect the deuterium nuclei of the solvent molecules (such as $CDCl_3$ or d_6-acetone). This signal is used for a field-frequency lock. A block diagram of such a typical spectrometer is shown in Figure 3.1 . For the simplest experiments many of the features given in Figure 3.1 are not necessary.

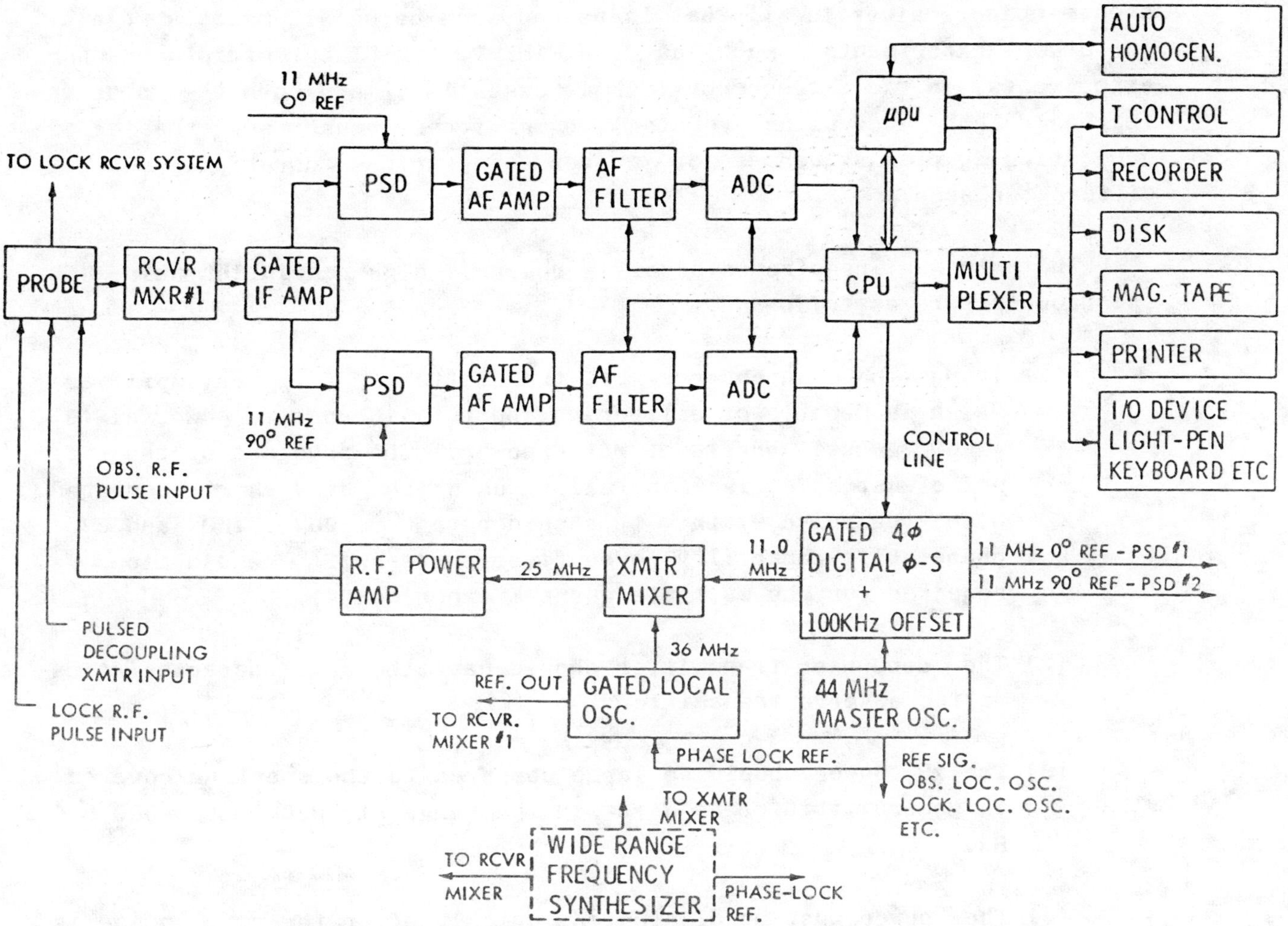

Figure 3.1 Block diagram for a modern pulsed NMR spectrometer. See text for details.

For example, if a proton NMR spectrum is needed, and if a superconducting magnet or stable permanent magnet is used, then a single 100 MHz transmitter-receiver is all that is needed when using a 2.3 tesla field. In other experiments, such as the INEPT, 2D-FT or multiple quantum experiments, a field-frequency lock is essential, and both the observe (carbon) and decoupling (proton) transmitters must be capable of generating complex sequences of pulses of various durations and of different phases.

The minimum requirements of a modern, high-resolution pulse NMR spectrometer are as follows:

(1) It must have an observe r.f. transmitter (r.f. power amplifier) which is capable of delivering short (1-10 μsec), powerful ($B_1 \geq 30$ gauss) bursts of r.f. power. The envelope of the r.f. pulse must have rise and fall times which are short compared with the pulse width. It should have 0°, 90°, 180° and 270° phase-shift capability (even finer phase-shift resolu-tion is required for the multiple quantum-experiments).

(2) The decoupler transmitter should have the same characteristics as the observe transmitter.

(3) The γB_1 power should be large compared to the spectrum covered. In general this means a γB_1 in the range of 10,000 to 500,000 Hz.

(4) The power must be dissipated quickly after the r.f. pulse is switched off.

(5) The sample must be efficiently coupled to the transmitter/ receiver coil such that the transmitter power is used effi-ciently and the nuclear signal is efficiently coupled to the receiver system.

(6) The receiver must recover very quickly (3-5 μsec or less) from any overloads generated by the application of the r.f. pulse.

(7) The receiver and transmitter should be well isolated from one another in order to realize minimum overload conditions and to achieve the fastest recovery times.

(8) The B_1 r.f. field must be as homogeneous as possible over the sample volume. The B_0 dc magnetic field should also be very homogeneous, as in any high-resolution NMR experiment.

(9) There should be minimal irradiation of r.f. energy at the resonance frequency into the laboratory in the absence of the γB_1 excitation pulse.

(10) The pulse programmer should have a very stable time base, operate under the control of a real-time clock (preferably phase locked to the r.f.frequency of the spectrometer) and have a time resolution of 0.1 μsec or better. At least four phases of 0°, 90°, 180° and 270° should be available. For multiple quantum experiments even finer phase resolution is required.

It is not possible in an introductory text such as this one to discuss in detail all, or even a substantial fraction, of the hundreds of pulse NMR experiments which are possible with a modern pulse NMR spectrometer. We discuss here only some of the more important considerations and indicate in general what features are needed.

The diagram in Figure 3.1 shows a broad overview of a modern pulse NMR spectrometer. In a 4.7 tesla (47,000 gauss) magnetic field, carbon-13 nuclei resonate at about 50 MHz, protons resonate at about 200 MHz and deuterium nuclei at about 30 MHz. All three of these frequencies are phaselocked to a common oscillator, and the operating frequencies of the three r.f. sources are controlled by the central computer. In some cases only the observe (carbon-13) and decouple (proton) frequencies can be varied. The deuterium lock frequency is fixed and the deuterium resonance condition for the lock sample is established by changing the magnetic field. The phase of the transmitter frequencies for both the observe and decouple channels are under computer control, and in most cases the amplitude of the decoupler power is also computer controlled.

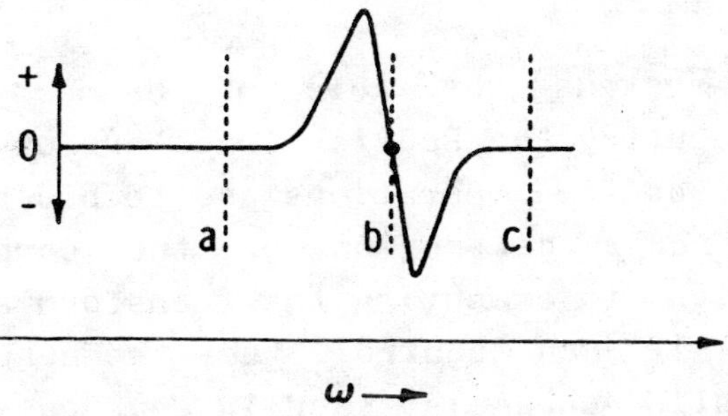

Fig. 3.2 Lock signal showing "error signal" amplitude vs. error frequency, $\Delta\omega$, where $\Delta\omega$ is the difference between the Larmor frequency ω_o and the Larmor field B_o.

The lock channel of the spectrometer stabilizes the system so that long-term signal averaging can be carried out. Its operation is basically rather simple. The solvents used for most ^{13}C NMR studies contain deuterium. Typical solvents are D_2O, d_6-acetone, $CDCl_3$, and d_6-dimethyl sulfoxide. They share a common property in that they all have single deuterium resonance signals. By proper adjustment of the spectrometer a "spectrum" similar to that shown in Figure 3.2 is obtained as the deuterium r.f. is swept. If the r.f. is adjusted so that this dispersion signal is at point (b) and the sweep then turned off, the signal will remain at (b). Any small variations in either the magnetic field or the spectrometer frequency will cause the system to deviate from point (b) and a positive or negative signal will register in the lock channel; the sign of the dc signal tells whether the system has drifted up or down in field (or frequency). This dc signal is used as an "error signal" and can correct very minute changes (up to one part in 10^9 or better) in the stability of the spectrometer system. The error signal is incorporated into a feedback system and compensates for any instabilities in the magnetic field or any one of the three radio frequencies. This system allows one to routinely obtain very high resolution (0.1 Hz or better) spectra even with time averaging up to 24 hours or more.

It is relatively common to "tune" the homogeneity of the B_o field by adjusting the Z1 and Z2 homogeneity controls for a maximum amplitude of the lock signal. This, however, is not a good way to obtain the optimum homogeneity, for at least two reasons: (1) deuterium resonance signals tend to be rather broad (especially the D_2O resonance line) and hence not very sensitive to the B_o field's homogeneity; (2) the deuterium gyromagnetic ratio is relatively small. A superior method of adjusting the magnetic field to a state of high homogeneity is to observe the FID signal of a sharp proton line and adjust Z1 and Z2 to give a maximum FID amplitude at a point about 1.0 sec after the beginning of the FID.

In early FT-NMR systems the role of the computer was minor and primarily limited to executing the Fourier transform on the FID signal, performing some data reduction operations and recording the spectrum on an x-y recorder. It is now rather common for the computer to play a major role. In addition to executing the transform and displaying the frequency domain data, it can control the frequency of the lock, observation, and decoupling channels, and in some cases the homogeneity of the magnetic field and the sample temperature. It is now common practice for a magnetic tape cassette, a floppy disk, or a large disk unit to be an integral part of the spectrometer system. This increases the flexibility of the system considerably since a large number of software programs as well as large amounts of data can be stored. If the computer has the capability of controlling the spectrometer, a large number of difficult experiments can be run in the automatic mode. Computer control also means that as new methods and techniques are developed the spectrometer's capability can be expanded, in many instances by writing new software, without the need to buy new hard-wired accessories.

In all modern spectrometers it is essential that the multiple r.f. phase capability mentioned earlier be included. This is required for the systematic noise reduction discussed later in this chapter as well as for the T_1, T_2 and $T_{1\rho}$ experiments described later. It is also absolutely essential for the new techniques being developed to execute high-resolution NMR studies in solids. In most cases the accuracy and stability of the relative phases is quite important. For operation in the quadrature detection mode, accurate adjustment of the 0° and 90° phases is important and it is essential that the phases be drift free. One way to generate accurate, stable phases is to use digital techniques. By using a pair of matched, digital binary dividers enclosed in the same integrated circuit (IC) device, it is possible to convert a single-phase r.f. signal into four separate digital signals that differ one from another in that their phase angles are 0°, 90°, 180°, and 270° (measured relative to a standerd frequency). The frequencies of these four signals are the same. The details are shown in Figure 3.3. A binary digital divider or "flip-flop" is a device that changes its output state from high (+5 V) to low (0 V) whenever it experiences a sharp rise at its input; no change in output is made when the input changes from a high state to a low state.

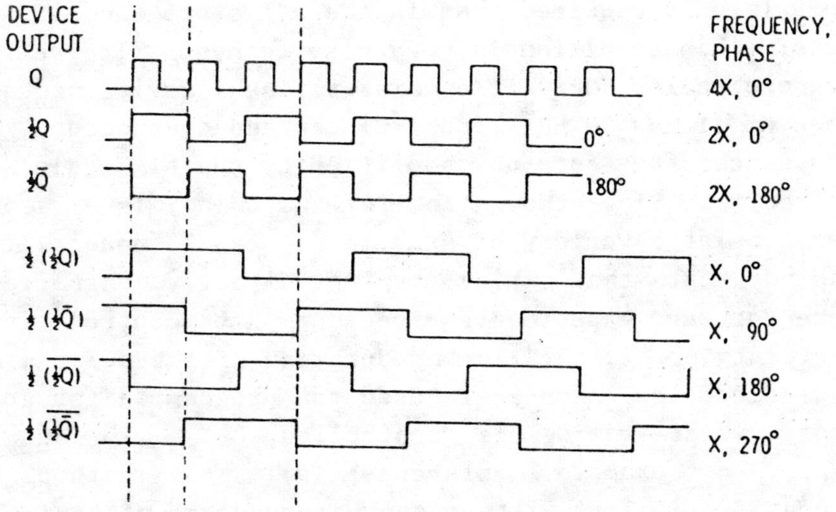

Figure 3.3 If a squarewave frequency, 4f, is connected to the input of a dual J-K flip-flop, the outputs will be four separate squarewaves each with frequency f, but with relative phases of 0°, 90°, 180° and 270°.

These devices are usually constructed such that two outputs are simultaneously available; the phases at these outputs differ by 180°. Thus when one output is high, the other one is low, and vice versa. As can be seen in Figure 3.3, if one starts with a digital frequency of 4f MHz, the final result is a four-channel output of frequency f MHz, with relative phases 0°, 90°, 180° and 270°. In order to ensure a high phase accuracy it is desirable to work at a reasonably low frequency; this is because the phase accuracy depends upon the ratio of the inverse frequency f^{-1} of the square wave to the rise time of the pulse. Since at 11 MHz, f^{-1} is about 10^{-7} sec and typical rise times of digital dividers are about 10^{-9} sec, the uncertainty in the phase angle is 1 percent or less. By using trimmer capacitors the accuracy may be adjusted to 0.1 percent or better.

By incorporating the r.f. phase shifting and the frequency offsetting at the i.f. (intermediate frequency) range it is relatively easy and inexpensive to operate at any frequency between a few megahertz and several hundred megahertz (required when superconducting magnets [SCM] are used) with the same accurate phase capability and offset capability. The spectrometer system shown in Figure 3.1 uses gated mixing and has the advantage that the gating requirements are now much easier to meet.

Most early FT-NMR spectrometers used frequency sources that operated at the Larmor frequency. If even a small amount of this frequency found its way to the receiver coil or the preamplifier via ground loops or other means, a number of unhappy consequences ensued. In the gated mixing scheme it is only required that in the off state the r.f. signals be 50 mV or less; this condition is very easy to meet. If the voltages on the mixer are below 50 mV it remains a linear device and no mixing occurs. Leakage of 11 and 39 MHz (for ^{13}C) can and does occur, but this does no harm since they are not amplified by the high-gain, high-Q-preamplifier circuits. Since the gating (or switching) is all controlled by the computer, a large variety of experiments can be done. It should perhaps be noted that the gating is not directly controlled by the computer, since for many experiments direct computer control introduces too much uncertainty or "jitter" in the duration of the pulse widths and their timing. Instead the computer is used to program a programmable, read-only memory whose timing is controlled by a 10 MHz crystal oscillator (see, for example, reference 19). This method has the advantage of great flexibility without the disadvantage of timing jitter. The gating feature is also important in the decoupling channel; examples of gated decoupling will be given later in this chapter.

Gating techniques are also used in the receiver circuitry of most modern systems. This is quite easy to do since most current, wideband, r.f. amplifier chips provide an automatic gain control capability. It is important to turn the signal amplifiers off when the transmitter pulse is on in order to prevent overload and ringing in the detection and filter circuitry. Another signifigant development is that of quadrature phase detection, which is discussed in more detail later in this chapter.

One distinct advantage of the sort of spectrometer shown in Figure 3.1 is that it can be used at any frequency from a few megahertz up to 600 MHz or even higher. This is done simply by replacing the gated local oscillater by a general-purpose frequency synthesizer. Since all the phaseshifting is done at the intermediate frequency no additional electronics for this purpose is required when the synthesizer is added. Since gated mixing is used, isolation and shielding problems of the synthesizer are greatly reduced. For operation at frequencies above 100 MHz, double-frequency conversion methods make the problems of eliminating unwanted side bands much simpler. (It is most difficult for this author to understand why most manufacturers use different spectrometers for different magnetic field strengths when a single well-designed system can do the job at any field equally well and at an overall reduced cost.) Similar spectrometer designs or conversions have been published by several other workers (19-22).

3.2 The Sample Probe

The sample probe is one of the most important parts of a NMR spectrometer, since in large part, it determines the sensitivity of the system and the quality of the spectra obtained. Optimization of the probe operating parameters is especially important when signal averaging is required, since an N-fold increase in sensitivity results in an N^2-fold decrease in the time required to obtain the spectrum. The modern probes now available allow one to observe a very wide range of nuclei (^{103}Rb to ^{31}P, which corresponds to a frequency range of 15 to 202 MHz for a field of about 12 tesla). The sensitivity of these new probes is one to two orders of magnitude better than their fixed frequency "equivalents" used five to ten years ago. It is relatively easy with these new probes to change the observation frequency rapidly. In special probes that are now available one can change from proton to carbon-13 observation by typing in a command at the computer console.

The design of the probe is very important when the amount, solubility and concentration of the sample are limiting factors. Since the beginning development of ^{13}C NMR, it has been popular to use rather large sample volumes. By using 20- or even 30-mm-diameter sample tubes, 10 to 15 ml or more of the sample can be used in the NMR sample tube. At such large tube diameters, however, great care must be used in order to maintain adequate r.f. power requirements. This is especially important if T_1 studies or quantitative analysis work is planned. In addition to the possible difficulties that may arise because of the probe design, there is also the problem, in the long run, of financing an experimental program that requires large amounts of costly deuterated solvents and biological samples. This is not to say there is no application for large samples. In many cases the samples in question are not very soluble or it is important to do experiments at relatively low concentrations (10^{-4} M or less). There are, however, many times when it is not possible or desirable to use such large sample tubes. The design and use of small sample tubes (1.6 mm o.d., 30-50 μl volume) has now made it possible to obtain good carbon-13 NMR spectra on as little as 100 μg of sample. Good proton spectra can be obtained with as little as 100 ng of sample. Such work requires a small (about 1.8 mm) receiver/transmitter coil, and very great care must be used in sample preparation.

3.3 Decoupling Considerations

For both ^{13}C and ^{15}N NMR studies it is necessary to decouple the fine structure arising from the spincoupling to the protons. It has been common practice to use proton "noise" decoupling in the standard ^{13}C or ^{15}N FT-NMR experiments (7). This method, however, is not too effective and, especially at higher fields, it becomes very difficult to decouple all the proton-induced fine structure in the ^{13}C spectrum and at the same time keep the r.f. heating effects under control. A much more effective method of decoupling has been described by Anet (23) and even more efficient decoupling techniques have been developed recently by a number of laboratories (24-29). From 1984 to 1986, with the advent of very greatly improved computer and r.f. electronics, a number of people have come up with very greatly improved decoupling schemes (30-32).

3.4 Gated Decoupling

In the normal $^{13}C-\{^1H\}$ or $^{15}N-\{^1H\}$ FT-NMR experiments the proton decoupler is operated continuously. In this mode of operation, the ^{13}C NMR spectrum consists of numerous single resonance lines with no proton fine structure and with the presence of the nuclear Overhauser enhancement (NOE). It is often helpful or even essential, however, to observe the proton-induced fine structure or to suppress the NOE. A variety of such experiments can be executed by turning the proton decoupler on and off (i.e., by "gating" the decoupler) at the proper times. Figure 3.4 shows some examples of the sorts of experiments that are now more or less standard.

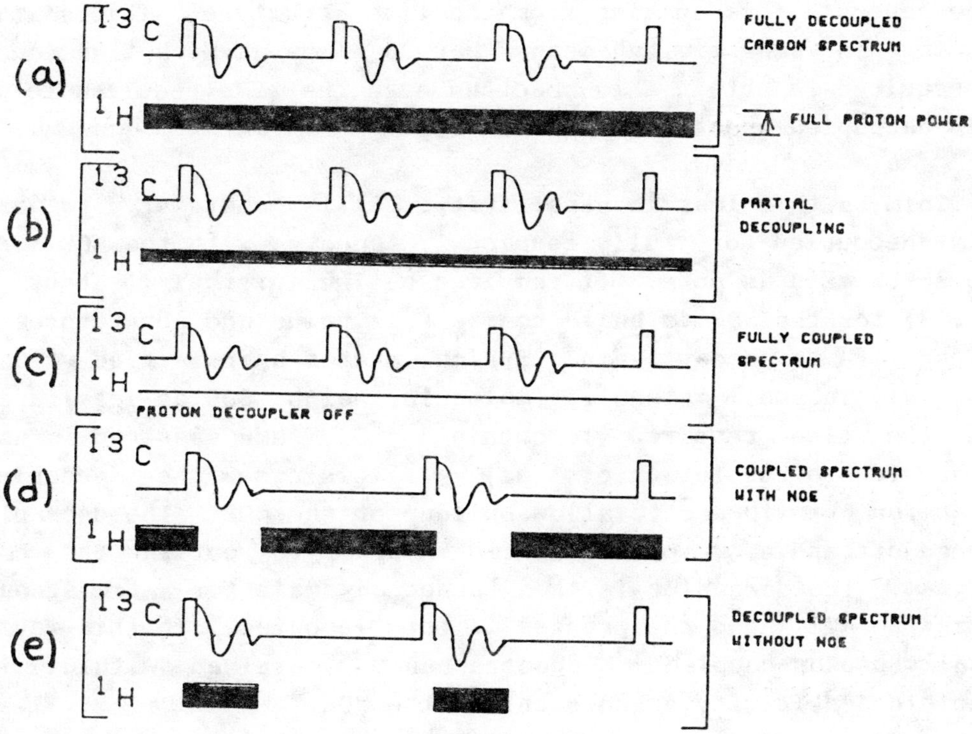

Figure 3.4 (a) Spectrometer operation of the "standard" ^{13}C observation with broad-band ^1H decoupling mode. This is denoted by $^{13}C-\{^1H\}$ and generally gives a full nuclear Overhauser enhancement (NOE). (b) Partial decoupling. The spin-spin splittings are all shown, but are greatly reduced in size; the full NOE is maintained. (c) Fully coupled ^{13}C spectrum as in (c), but with the NOE. (e) Decoupled spectrum as in (a), but without the NOE.

Figure 3.4a depicts the standard FT-NMR experiment, in which the ^{1}H irradiation power is on continuously, ^{13}C pulses are applied repetitively, and the corresponding FID signals are collected and added together in the computer. In Figure 3.4b the proton decoupling power is substantially reduced; the result is that the proton-induced fine structure is present, but the observed $^{13}C-^{1}H$ coupling constants are all reduced. The experiment is still very useful because it can show how many protons are bonded to each carbon atom. For example, a single ^{13}C doublet indicates one proton, a triplet shows two protons, and a quartet indicates three directly bonded protons. In the case of partial decoupling the full NOE is maintained. In Figure 3.4c the proton decoupler is turned completely off. In this case the full $^{13}C-^{1}H$ coupling is seen and there is no NOE. It should be noted that this is a very time-consuming experiment since the NOE is lost (a factor of three in sensitivity) and the decoupled ^{13}C singlet signals are now divided among the several lines arising from the fine structure. This sixfold or greater loss in "sensitivity" means that 36 times as much time (or more) will be required to obtain this spectrum over the time required to obtain the fully decoupled spectrum.

To minimize this loss in sensitivity, it is customary to execute partially decoupled or fully coupled ^{13}C spectra with the NOE. Figure 3.4d shows how this is done. It requires a time interval of about three times $T_1(^{1}H)$ for the NOE to build to its full value and about three times $T_1(^{1}H)$ for it to decay again. Typical proton spin-lattice relaxation times $[T_1(^{1}H)]$ in small molecules (molecular weight 600 or less) are 2-5 sec and the time required to obtain a ^{13}C FT-NMR spectrum is usually about 1 sec (for a resolution of 1 Hz). In this case the decoupler is turned on for 5 to 10 sec to allow buildup of the NOE. The decoupler is then turned off and a ^{13}C pulse applied in order to obtain the coupled ^{13}C spectrum. The decoupler is then turned on again for a few seconds to revitalize the NOE, and the process is then repeated. In this way fully or partially proton-coupled ^{13}C spectra can be obtained without losing the threefold sensitivity enhancement of the NOE.

The converse of this experiment is shown in Figure 3.4e, a decoupled ^{13}C spectrum without NOE. Suppression of the NOE is done primarily for two reasons. In one case ^{13}C spectra are recorded both with and without NOE (Figure 3.4a and e, respectively) in order to measure the actual NOE. If the ^{13}C nuclides are relaxed entirely by the dipole-dipole interaction with the protons (see references 7 and 10), then a sensitivity enhancement of 2.9 will be observed. If this is not observed it indicates that "other" relaxation mechanisms are contributing. In most cases the mechanisms are spinrotation (for small molecules) or paramagnetic impurities (such as dissolved oxygen, traces of iron, or copper ions). At the higher magnetic fields (6-12 tesla) that are becoming common today, chemical shift anisotropy (CSA) contributes significantly to relaxation and diminishes the NOE.

It is now becoming more common to make quantitative ^{13}C measurements in order to determine the relative amounts of the different sorts of carbon atoms present. This is sometimes a tricky task because of the differences in spin-lattice relaxation times (T_1) and NOE values for the various carbon nuclides. The T_1 distribution problem can be solved either by using a viscous solvent or by lowering the temperature to the point where the ^{13}C T_1 values are all reasonably short and about equal (about 1 sec). In any event one must wait about five times the longest T_1 between pulses. The NOE distribution problem can be solved by performing the experiment shown in Figure 3.4e. The decoupler is turned on for only the 1 sec, during which spectra are recorded. One then waits several seconds with the decoupler turned off (in order for any small initial NOE buildup to decay), and the experiment is then repeated. When it is possible, both the T_1 and NOE distribution problems can be circumvented by adding "shiftless" relaxation agents such as $Cr(AcAc)_3$ to the solution. This shortens the relaxation times to about 1 sec and suppresses the NOE. The loss in NOE enhancement is usually more than compensated for by the fact that the relaxation times are now all relatively short and large ^{13}C pulse angles can be used.

If one executes the experiment in such a way that T_1 and NOE differences are eliminated or accounted for, then it is rather routine to carry out quite accurate quantitative measurements, with results that agree very well with other analytical methods. It is very important to maintain a high S/N ratio and to use an adequate number of data points for quantitative work. Sampling theory requires that if a spectral range of Δ Hz is to be observed, then a sampling rate of 2Δ Hz must be used. For ^{13}C observation at 50 MHz, Δ is typically 10 kHz. For a resolution of 1 Hz it is necessary to sample for 1 sec . This means that 20,000 data points are required, but because of the 2^q requirement imposed by the Cooley-Tukey algorithm one usually samples for a shorter time in order to limit the number of data points to 16K (16 x 1024 = 16,384). After execution of the Fourier transformation one is left with 8K data points for the absorption spectrum and 8K data points for the dispersion spectrum, less than one data point per hertz. For quantitative work, this is clearly not adequate, unless rather broad lines are involved, such as in viscous polymers or heavy oils (note that increased viscosity decreases molecular motions and results in shorter relaxation times). For molecules with molecular weights up to about 600 or 700, ^{13}C line widths are usually 1 Hz or less. Since four or more data points per line are needed for accurate intensity measurements, one must consider the use of 64K transforms or partial spectra. It should be mentioned that really accurate T_1 measurements require accurate total line intensity information.

Two methods currently used to increase the number of data points per spectral line are exponential filtering and zero filling. If the resonance lines in the spectrum are not closely spaced, then a large amount of exponential filtering can be added by the computer. This will greatly enhance the S/N ratio and broadens the lines appreciably. The result is that without adding to the total number of data points, it greatly increases the number of data points per line with a concomitant increase in the accuracy of T_1 measurements and quantitative measurements. An alternative method is to use zero filling. If x data points are taken, one may add an additional x data points, which are all zero. For example, if 16K zeros are added to 16K original data points, one can execute a 32K transformation that results in 16K real (absorption) and 16K imaginary (dispersion) data points in the frequency domain. Thus, zero filling has the advantage of doubling the number of data points per line. It also has the distinct advantage of increasing the resolution by a factor of two. It has been argued that this is not possible since adding zeros does not add new information and consequently cannot improve the resolution, and that even the gain in spectral definition (as opposed to resolution) is a bit of a mathematical boondoggle. This point of view, however, is somewhat extreme; it can be reasoned that the Fourier transformation itself adds no new information. This is in fact true since all the information is contained in the FID, but it is not in a very usable form. Adding zeros does not add new information but it does allow one to make full use of the information contained in the FID. As a result of the Fourier transformation, half of the resolution information contained in the FID is lost; adding the additional zeros allows one to regain this misplaced information. Adding more than x zeros to the original x data points serves no useful purpose, other than giving more interpolated data points between each of the original data points. Another way to improve the resolution is to record 16K rather than 8K data points. This improves the resolution twofold, but doubles the scan time and degrades the S/N by a factor of $2^{1/2}$ for an equal number of scans.

3.5 Computer Control and Data Acquisition

Most of the experiments done with modern pulse NMR spectrometers require a large number of r.f. pulses with carefully defined widths, phases and timing intervals. Great demands are also placed on both channels of the receiver circuitry (see Figure 3.1). In all modern spectrometers quadrature detection is used because of the increased transmitter efficiency (two to three times more effective use of transmitter power), and the S/N ratio is improved by 40 percent. These very substantial gains are obtained at the cost of considerably more complex receiver channels. For quadrature detection two receiver channels, A and B, are required and the signals they detect are 90° out of phase with one another (see Figure 3.1). In order to obtain accurate, meaningful data it is essential that the gain and phase stability of the circuits be very high or error correction schemes must be implemented. Since it is impossible to build instruments with the required absolute stability, error compensation methods are almost always used.

Consider the process of obtaining a simple proton NMR spectrum. Usually the A channel uses a fixed 0° reference phase and the B channel uses a fixed 90° reference phase. These relative phases can be changed under computer control, but for this experiment they are usually held constant. The transmitter phases are varied in an eight-pulse sequence: 0°, 0°, 180°, 180°, 90°, 90°, 270°, 270°. When the phase, Φ_{tx}, of the transmitter voltage is zero, it is in phase with the A channel of the receiver, and consequently for $\Phi_{tx} = 0^o$, the A channel signal will be an in-phase or dispersion mode signal. The B channel is 90° out of phase when $\Phi_{tx} = 0^o$ and thus records an absorption mode signal. We shall store the A channel signals in computer even memory locations beginning at address 0 and the B signals in odd memory locations beginning at address 1. When $\Phi_{tx} = 0$ we add the A signal to the even addresses and we add the B signal to the odd addresses (see Figure 3.5).

When the transmitter phase is shifted to 180° ($\Phi_{tx} = 180^o$) for pulses 3 and 4 and the receiver phases remain constant ($\Phi^A_{rx} = 0^o$, $\Phi^B_{rx} = 90^o$), the signal now received at A is an inverted dispersion mode signal and at B an inverted absorption mode signal (see Figure 3.5). These signals are subtracted from the even and odd memory locations for signals A and B, respectively. For pulses 5 and 6 in the eight-pulse sequence we change the transmitter phase to 90° ($\Phi_{tx} = 90^o$). Channel A now receives a negative dispersion mode signal and B receives a positive absorption mode signal. In this case the channel A signal is subtracted from the odd memory locations and the B signal is added to the even memory locations. Finally, for Φ_{tx} = 270° (= -90°), the B signal is a negative dispersion signal and A is a positive absorption signal. In this case, A is added to the odd memory locations and B is subtracted from the even memory locations (see Figure 3.5).

Φ_{tx}	Output A	Output B	Computer memory even	odd	Φ_{rx}
0°	+Disp	+Abs	+A	+B	0°
90°	-Abs	+Disp	+B	-A	0°
180°	-Disp	-Abs	-A	-B	0°
270°	+Abs	-Disp	-B	+A	0°

Φ_{tx} for eight-sequence

pulse #	1	2	3	4	5	6	7	8
Φ_{tx}	0	0	180	180	90	90	270	270

Figure 3.5

For this sequence of eight pulses, the even memory locations contain a positive dispersion mode signal with about eight times the amplitude of the single scan signal. Similarly, the odd memory locations contain a positive absorption mode signal. Any coherent noise present is almost entirely canceled and all signals are processed half the time through channel A and half the time through channel B. The result is that any changes in receiver gain in either channel will be compensated. Compensation schemes such as this one (and even more complex ones) are all easily executed by the system's computer. Other equivalent schemes are possible which involve changing the phase of the reference signals for receiver channels A and B.

In more complex pulse experiments such as INEPT and ERNST, which are discussed in detail in Chapter 6, 16-pulse phase cycles are used. In multiple quantum experiments, it is common to use 64-(or more) pulse phase cycles to correct for various experimental errors and artifacts. For more details on the hardware and software requirements of complex experiments, the reader is referred to a number of the recent excellent treatises on this subject (33-35).

CHAPTER FOUR

RELAXATION MECHANISMS

In this chapter we draw a few simple physical pictures which we hope will give some insight into the mechanisms by which a system of nuclear spins relaxes when put into an intense magnetic field or when perturbed from its equilibrium state. That is, how does the spin system come to equilibrium with its environment, "the lattice"? We show that the Fourier component of the random motion at the nuclear resonance frequency ω_0 is responsible for relaxation and that the magnitude of this component along with the strength of the interaction which couples the nuclear precession to the molecular motions jointly determine the values of T_1 and T_2. Finally, we use these results to sketch the development of equations which quantitatively relate T_1 and T_2 to dipole-dipole, spin rotation, and other interactions, and we give some examples of what sorts of useful chemical information and structural information can be obtained from such data.

4.1 Frequency Distribution of Molecular Motions

The frequencies of molecular motions in a liquid vary over a very wide range. At any given time some molecules are moving quite slowly, while others are moving very rapidly indeed. Another way to characterize these motions is by the time a "typical" molecule remains in any given position. On the average, in a nonviscous liquid a molecule remains in one state of motion for about 1 psec (10^{-12} sec). After this time it suffers a collision which changes its state of motion. If the molecule persists in some state of motion for a time of 1 psec, we expect that its motion will have frequency components from 0 Hz to about 10^{12} Hz. This is closely analogous to the pulsed or switched oscillator which was discussed in Chapter 3. When a sine wave of frequency f_c is turned on for a short time, say τ seconds, we obtain not only a component at frequency f_c, but also components at frequencies over a range of frequencies of approximately ($f_c \mp \tau^{-1}$) Hz.

The intensity of the motional frequency components at the nuclear resonance frequency, at twice this frequency, and at zero frequency are important in NMR relaxation. Recall from Chapter 2 that random motions of the small nuclear magnetic moments in neighboring molecules are responsible for both spin-lattice (T_1) and spin-spin (T_2) relaxation. Both slow and fast motions are effective for spin-spin relaxation, but that only the fast motions, near the Larmor frequency, are effective in spin-lattice relaxation. Motions at about twice the Larmor frequency are also effective in both spin-spin and spin-lattice relaxation. The contributions to T_1 and T_2 from motional frequencies present at twice the resonance frequency arise from a sort of molecular Doppler effect. This is discussed in detail by Andrew (36).

In order to discuss in a more quantitative way how the relaxation times T_1 and T_2 are related to the molecular motions, we need to introduce some new parameters. Let τ_c be the "average" time between collisions for a molecule in some state of motion. If the value of τ_c is such that large Fourier components of the molecular motion exist at the resonance frequency, ω_o, then we expect relaxation to be most efficient. Consequently, we expect the relaxation times to be minimized. If, however, τ_c is too long or too short, the Fourier components at ω_o are much smaller, and much longer values for T_1 and T_2 result. This is in fact the case, as was shown by Bloembergen et al. (37). Their results are shown in Figure 4.1. For a sample of glycerine, they plotted the dependence of T_1 (proton) on the molecular correlation time τ_c at two different resonance frequencies. In the high-temperature region, to the left of the T_1 minima on the graph, T_1 is frequency independent, and it decreases with decreasing temperature. This behavior is typical of any molecule which is characterized by τ_c values where $\omega_o\tau_c \ll 1$. This is generally the case for small molecules in the liquid state. At both resonance frequencies ($v_o = \omega_o/2\pi$ = 29 and 4.8 MHz), T_1 goes through a minimum and the value of T_1 at the minimum is proportional to ω_o. As we shall see later, this behavior occurs whenever $\omega_o\tau_c \simeq 1$, a situation that obtains in large polymer molecules, in large organic or biochemical molecules or in samples, such as glycerine, which have been cooled to a highly viscous or glassy state. In the low-temperature region of Figure 4.1, to the right of the T_1 minima, T_1 is frequency dependent and its value increases with decreasing temperature. This situation is found whenever $\omega_o\tau_c \gg 1$.

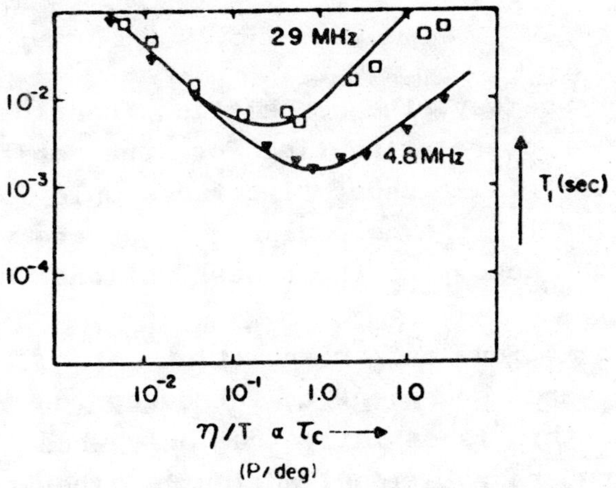

Figure 4.1 The relaxation time, T_1, is plotted as a function of the viscosity for a sample of glycerine. As the viscosity is increased (by decreasing the temperature), the relaxation time decreases until it goes through a minimum at about $\omega_o\tau_c = 1$. 29 MHz data are indicated by the open squares and 4.8 MHz data are shown by the solid inverted triangles. See reference 37 for details.

Before we can place our understanding of these experimental observations on a firmer, more quantitative foundation, it is necessary to introduce two more parameters which are closely related to one another, namely, a correlation function $K(\tau)$ and its Fourier transform, the spectral density function $J(\omega)$. Since these two parameters form a Fourier transform pair one may write

$$J(\omega) = \int_{-\infty}^{+\infty} K(\tau)\exp(i\omega\tau)\,d\tau. \qquad (4.1)$$

We shall use $K(\tau)$ to express τ_c in a more quantitative fashion, and we use $J(\omega)$ to give the intensities or probabilities of the molecular motions at the resonance frequency, ω_o.

Now $K(\tau)$ is a function which is rather similar to the FID signal observed in a pulse NMR experiment. The FID signal decays more or less exponentially with a time constant T_2^* as discussed in Chapter 2. The time constant which characterizes the decay of the FID signal (be it T_2^* or T_2) is sometimes referred to as the <u>phase memory time</u>, since it is related to the time for M_{xy} to lose phase coherence and decay to zero. A rapidly decaying FID signal arises from a sample which has a short phase memory time.

In exactly the same way, the correlation function $K(\tau)$ is related to a phase memory time or correlation time for the molecular motions. If $K(\tau)$ decays to zero in a short time the sample is characterized by a short correlation time τ_c, the molecular motions are very rapid, and the molecules have a short memory of their past motional history.

To describe the motion and/or position of a rigid body, it is customary, helpful, and convenient to introduce functions related to the spherical harmonics (this is strictly true only when the motions about which we are speaking are rotational motions). They are defined as:

$$Y_0 = r^{-3}(1 - 3\cos^2\theta),$$

$$Y_1 = r^{-3}\sin\theta\cos\theta\exp(i\Phi), \qquad (4.2)$$

$$Y_2 = r^{-3}\sin^2\theta\exp(2i\Phi),$$

where the r, θ, and Φ are the usual spherical coordinates. As the molecule moves, r, θ, and Φ become functions of time, and hence the Y_i (i=0,1,2) given in equation 4.2 become functions of time. Since the relaxation times T_1 and T_2 are related to the <u>average</u> behavior of a collection or ensemble of nuclei, one is particularly interested in the average way in which the nuclei move about. The correlation functions $K_0(\tau)$, $K_1(\tau)$, and $K_2(\tau)$ give this information. They are defined as

$$K_i(\tau) = \overline{Y_i(t)Y_i^*(t + \tau)}, \qquad i = 0,1,2, \qquad (4.3)$$

where Y_i^* is the complex conjugate of Y_i, and the bar over the product indicates an average over the ensemble of nuclei, or an <u>ensemble average</u>. It can be shown (reference 38, chapter 9) that it is quite reasonable to let each of the $K_i(\tau)$ have the functional form

$$K_i(\tau) = K_i(0)\exp(-|\tau|/\tau_c), \qquad (4.4)$$

where

$$K_i(0) = \overline{Y_i(t)^2} = \overline{Y_i^2} = \int_0^{2\pi}\int_0^{\pi} Y_i^2 \sin\theta\, d\theta\, d\Phi. \qquad (4.5)$$

Straightforward integration yields the results

$$K_0(0)=(12/15)r^{-6}, \quad K_1(0)=(2/15)r^{-6}, \quad K_2(0)=(8/15)r^{-6}. \qquad (4.6)$$

Equation 4.4 shows in more rigorous terms what we have already discussed qualitatively, namely, that in a system characterized by a short correlation time τ_c, the molecules lose phase coherence, or memory of an earlier position, in a short time.

Just as one can investigate the frequencies present and their corresponding intensities, in a FID signal via Fourier transformation one can determine the motional frequencies and intensities present in the correlation functions. That is,

$$J(\omega) = \int_{-\infty}^{+\infty} K(\tau)\exp(i\omega\tau)\, d\tau$$

$$= \int_{-\infty}^{+\infty} K_i(0)\exp(-|\tau|/\tau_c)\exp(i\omega\tau)\, d\tau. \qquad (4.7)$$

By substituting equation 4.6 into equation 4.7 and integrating, we obtain

$$J_0(\omega) = (24/15r^6)[\tau_c/(1+\omega^2\tau_c^2)],$$

$$J_1(\omega) = (4/15r^6)\ [\tau_c/(1+\omega^2\tau_c^2)], \qquad (4.8)$$

$$J_2(\omega) = (16/15r^6)[\tau_c/(1+\omega^2\tau_c^2)].$$

These are the spectral density functions used in standard texts (10, 12) to relate reorientational molecular motions to the relaxation times T_1 and T_2.

By considering the dependence of these spectral density functions on the frequency for several values of τ_c we can gain a little more insight into the relation between T_1 and τ_c. In Figure 4.2 we have indicated how the typical function $J_0(\omega)$, given in equations 4.8, varies with frequency for short ($\omega_0\tau_c \ll 1$), long ($\omega_0\tau_c \gg 1$), and intermediate ($\omega_0\tau_c \simeq 1$) values of τ_c. The functional form and behavior of $J_1(\omega)$ and $J_2(\omega)$ is similar to that of $J_0(\omega)$. The fact that the area under these curves is constant [that is, $\int J_0(\omega)\, d\omega = \pi/2 \neq f(\tau_c)$] is important. It tells us that the amount of molecular power available for molecular motions is fixed and that changing τ_c changes only the way that the power is distributed. If τ_c is short, the molecules motions are distributed over a very wide frequency range (about 0 to τ_c^{-1} Hz) and all components of frequencies in this range have small but equal probabilities of being found in a typical molecule.

If τ_c is long, components with very low motional frequencies have a very high probability of occurring, but higher frequencies will be almost totally absent. In either case the Fourier component at the Larmor frequency will be small and T_1 will be long. In the intermediate region where $\omega_0\tau_c \simeq 1$, the motional components at ω_0 will be largest.

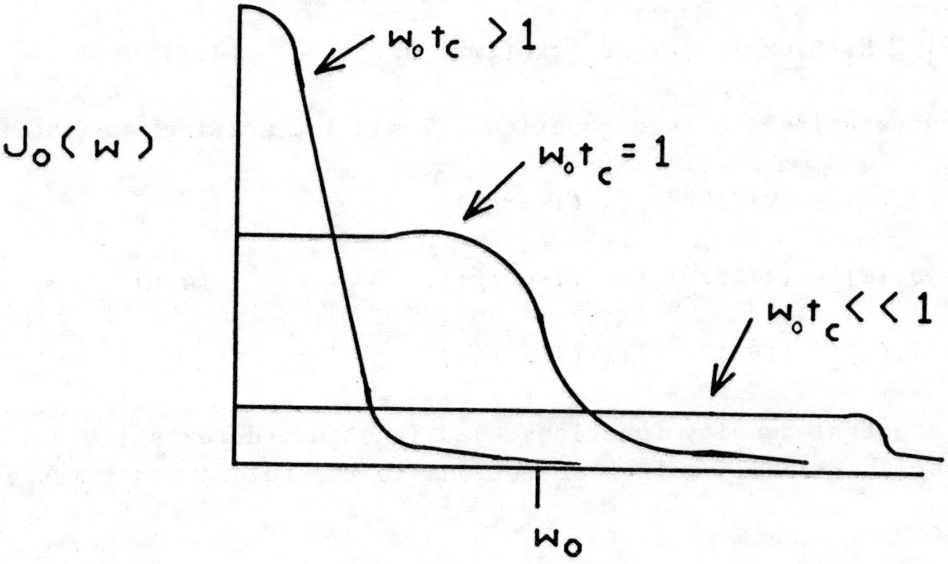

Figure 4.2 A plot of the spectral density function $J_0(\omega)$ as a function of the Larmor frequency, ω_0, for three different regions of the frequency-correlation time product. For $\omega\tau_c > 1$ the spectral density function is relatively large, but only over a small range of low frequencies. For $\omega\tau_c \ll 1$, $J_0(\omega)$ is small, but extends over a very wide frequency range. For $\omega\tau_c \approx 1$, $J_0(\omega)$ has an intermediate value over an intermediate range of frequencies. The relaxation time depends upon the value of J_0 at the Larmor frequency, ω_0. For the "extreme narrowing" condition where $\omega\tau_c \ll 1$, the relaxation time is not expected to depend upon the operating frequency, since J_0 is constant over a very wide frequency range.

4.2 Spin-Lattice Interactions

As we commented earlier, T_1 depends not only on the value of $J(\omega_o)$, but also on the strength of the coupling between the spin system and the lattice. A number of different physical interactions have been found to be important in coupling the nuclei to the lattice and hence in providing a link through which energy between these two systems can be exchanged. These processes are mediated by the following:

1) magnetic dipole-dipole interaction
2) electric quadrupole interaction
3) chemical shift anisotropy interaction
4) scalar coupling interaction
5) spin-rotation interaction
6) a combination of the above

In general, any mechanism which gives rise to fluctuating magnetic fields at a nucleus can produce relaxation. We expect, then, that one should be able to write

$$1/T_1 = R_1 = E_c^2 \, f(\tau_c), \tag{4.9}$$

where E_c represents the magnitude of the particular relaxation interaction and τ_c is, again, the molecular correlation time. If we know or can measure any two of the parameters given in equation 4.9, then we can calculate the third. As we shall see, this makes T_1 measurements very useful for many applications.

Since in many cases the dipole-dipole interaction is the dominant relaxation mechanism, we discuss it first and in some detail. The arguments which lead to the relaxation equations for the other interactions are quite similar to those for the dipole-dipole mechanism. Consequently, in these latter cases we give only the final equations, discuss some of the ramifications, and give a few examples of how they may be used to provide useful chemical information.

4.3 Dipole-Dipole Relaxation

Let us now direct our attention to two particular nuclei in a spin system, and let us stipulate that these two nuclei, which we label I and S, are contained in the same molecule. If the molecule is stationary, as it is, for example, in a solid, then nucleus I will see a total magnetic field, B_t, which is the sum of two parts: (1) a contribution B_o from the laboratory magnet, and (2) a contribution B_{loc} from the magnetic moment μ_S of nucleus S. Now it may be shown (12) that the strength of the interaction between two magnets depends on the magnetic moments of the two magnets, the separation of the magnets, and their relative orientation. Using the notation of Figure 4.3, the approximate field is:

$$B_t = B_o + B_{loc}$$

and

$$B_{loc} = \pm (\mu_S/r^3)(3\cos^2\theta - 1)(r/r), \qquad (4.10)$$

where r is the distance between nuclei I and S, and θ is the angle between the internuclear vector and the B_o magnetic field.

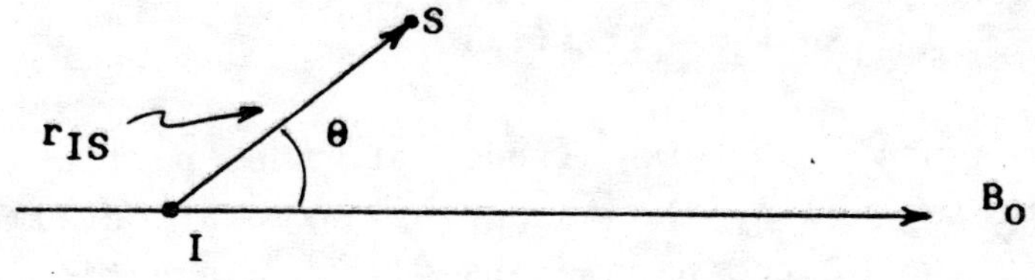

Figure 4.3 The local magnetic field experienced by nucleus I due to a neighboring nucleus S depends upon the third power of the distance between them and the angle which their internuclear vector makes with the applied dc magnetic field, B_o.

The $\pm$ sign arises because, depending upon the spin state of nucleus S, B_{loc} can add to or subtract from the contribution of the dc magnetic field B_0. For protons, a 20 gauss field amplitude, which is typical for water molecules in hydrated crystals, is equivalent to a separation of 80,000 Hz. Thus we see that this interaction can be rather large. As the molecule begins to move, the angle θ becomes time-dependent, and there are two consequences. First, if the motion is rapid enough, the average value of the local field goes to zero since

$$\frac{\int_0^{2\pi}\int_0^{\pi} (3\cos^2\theta - 1)\sin\theta \, d\theta \, d\Phi}{\int_0^{2\pi}\int_0^{\pi} \sin\theta \, d\theta \, d\Phi} = 0 \qquad (4.11)$$

and the resonance line becomes much narrower. By "rapid enough" we mean that the frequency of reorientation is fast compared with the width of the spectrum. In the case of the hydrated crystals cited above, water molecules must reorient much faster than 80 kHz in order to average these local field effects to zero.

The second, and for our present interests more important, consequence of this motion is that the nuclei I and S move about in space. As observed previously, it is this motion which causes the angle θ (or the function Y_0 given in equation 4.2), and consequently the local magnetic field given by equation 4.10, to fluctuate in time.

To derive specific equations for dipolar interactions we must consider the Hamiltonian operator associated with this particular interaction. Now, all of the relaxation mechanisms have associated interaction Hamiltonians which have the general form

$$\mathbf{H}_C = h\, \mathbf{I} \cdot \mathbf{T}_C \cdot \mathbf{O}. \qquad (4.12)$$

This represents the interaction of the nuclear spin I with another physical quantity, represented by the operator **O**. For example, for the interaction of I with another spin, **O** = **S**. For the interaction of I with the magnetic field, **O** = $\mathbf{B}_0$, and so on.

The tensor $\mathbf{T}_C$ represents the coupling interaction between I and **O**. In general, one can calculate the nine components, t_{ij}, of the tensor T in a <u>molecular</u> coordinate system, since in this frame of reference the t_{ij} are well-defined constants. In an NMR experiment, however, we are concerned with the orientation of the nuclear spin with respect to the <u>laboratory</u> coordinate system, that is, with respect to a set of axes which are fixed in the laboratory. Consequently needs to express the t_{ij} in terms of the lab system. This is a straightforward task since the $t_{ij}(lab)$ in the laboratory system are just linear combinations of the $t_{ij}(molec)$ in the molecular system.

Thus, if the molecule moves about rapidly, the $t_{ij}(l)$ become time dependent and we have a relaxation mechanism. The manner and the rate at which the $t_{ij}(l)$ fluctuate in time is usually expressed by lattice parameters, $Y_i(t)$, which are related to the spherical harmonics.

For our simple two-spin system (I and S), then, R_1 [$(1/T_1) = R_1$] for nucleus I will depend upon:

1) the strength of the coupling between the nuclei and the lattice (and thus will depend upon γ_I, γ_S, I, S, and r^{-3}) and

2) the frequency of molecular motions, or rather, the magnitudes of the spectral density functions, J_i, at the Larmor frequencies of the nuclei which are interacting.

The rate at which nuclei return to equilibrium, if they are put into a nonequilibrium configuration, is given by

$$dI_z/dt = I_z = -R_1(II)(I_z - I_o) - R_1(IS)(S_z - S_o), \qquad (4.13a)$$

$$dS_z/dt = S_z = -R_1(SI)(I_z - I_o) - R_1(SS)(S_z - S_o), \qquad (4.13b)$$

where

$$R_1(II) = \gamma_I^2 \gamma_S^2 \hbar^2 S(S+1) \times \qquad (4.14a)$$

$$[1/12\ J_o(\omega_-) + 3/2\ J_1(\omega_I) + 3/4\ J_2(\omega_+)],$$

$$R_1(IS) = \gamma_I^2 \gamma_S^2 \hbar^2 S(S+1) \times [1/2\ J_o(\omega_-) + 3/4\ J_2(\omega_+)], \qquad (4.14b)$$

where $\omega_- = \omega_I - \omega_S$, $\omega_+ = \omega_I + \omega_S$,

and the J_i's are those given in equations 4.8 above. The expressions for $R_1(SI)$ and $R_1(SS)$ are obtained by the obvious interchange of the labels I and S.

These equations assume that the relaxation is due only to intramolecular rotational motions (hence we can use the spherical harmonic functions, the Y_i's) and that the dipole-dipole interaction is the only relaxation process which is active. Since the two differential equations are coupled, we might expect to find that the recovery of equilibrium cannot be described by a single exponential function of time. In practice, the relaxation of simple, two-spin dipolar coupled systems has almost always been found to be described reasonably accurately by a single exponential function. This indicates that in most cases the cross-relaxation term, the $R_1(IS)$ term, is small compared to the $R_1(II)$ term. If another relaxation mechanism, such as relaxation via chemical shift anisotropy, also makes an appreciable contribution to the relaxation, then the relaxation can become very complex indeed (39).

Substitution of the expressions for the J_i's in equations 4.8 gives an expression for the relaxation rates as a function of the molecular correlation time. Thus

$$R_1(I) = \gamma_I^2\gamma_S^2 \lambda^2 S(S+1)r^{-6} \times$$

$$2/15 \left[\frac{\tau_c}{1+\omega_-^2\tau_c^2} + \frac{3\,\tau_c}{1+\omega^2\tau_c^2} + \frac{6\,\tau_c}{1+\omega_+^2\tau_c^2}\right] . \qquad (4.15)$$

In a similar fashion we may write an expression for R_2^I:

$$R_2(I) = \gamma_I^2\gamma_S^2 \lambda^2 S(S+1)r^{-6} \times \qquad (4.16)$$

$$1/15 \left[4\,\tau_c + \frac{\tau_c}{1+\omega_-^2\tau_c^2} + \frac{3\,\tau_c}{1+\omega_I^2\tau_c^2} + \frac{6\,\tau_c}{1+\omega_S^2\tau_c^2} + \frac{6\,\tau_c}{1+\omega_+^2\tau_c^2}\right] .$$

Thus we see that our qualitative pictures of the molecular motions which contribute to R_1 and R_2 are indeed correct: R_2 contains a term $J_o(0)$ at zero frequency, whereas R_1 contains terms only in the range of the resonance frequencies, and at sums and differences of them.

For rotational motion, which is all that is necessary to consider for <u>intramolecular</u> dipole-dipole interactions, the values of the $J_i(\omega)$ given in equations 4.8 are valid. For a system of like spins, such as in a water molecule, the expressions for R_1 and R_2 are (10):

$$R_1)_{rot} = 2/5 \quad \gamma^4\hbar^2 I(I+1)r^{-6} \left[\frac{\tau_c}{1+\omega^2\tau_c^2} + \frac{4\tau_c}{1+4\omega^2\tau_c^2}\right], \tag{4.17}$$

$$R_2)_{rot} = 1/5 \ \gamma^4\hbar^2 I(I+1)r^{-6} \left[3\tau_c + \frac{5\tau_c}{1+\omega^2\tau_c^2} + \frac{2\tau_c}{1+4\omega^2\tau_c^2}\right], \tag{4.18}$$

In mobile (i.e., nonviscous) liquids where the relation $\omega_o\tau_c \ll 1$ holds, R_1 and R_2 become equal and are given by

$$R_1 = R_2 = \quad 2\gamma^4\hbar^2 I(I+1)r^{-6} \ \tau_c. \tag{4.19}$$

For this situation, often referred to as the extreme narrowing condition, that R_1 and R_2 are frequency independent. On the other hand, when $\omega_o\tau_c \gg 1$ we see from equations 4.17 and 4.18 that both R_1 and R_2 are frequency dependent. This explains the frequency dependence of T_1 (proton) observed for glycerine at low temperatures (see Figure 4.1).

As we have observed, the above equations have been developed for rotational motion and consequently apply to intramolecular relaxation. The motions involved in intermolecular relaxation are primarily translational motions. These motions have different correlation times and lead to somewhat different expressions for the contribution of dipolar relaxation to T_1. For a system of two identical spins the translational relaxation rate is given by

$$R_1(\text{transl}) = 3/2 \quad \gamma^4\hbar^2 I(I+1)[J_1(\omega) + J_2(\omega)]. \tag{4.20}$$

This effect is of great importance when proton-containing solvents are used or when dissolved paramagnetic impurities are present. Since γ_e, the electron gyromagnetic ratio, is 657 times greater than γ_H it is easy to see how very small amounts of paramagnetic impurities such as O_2, Cu^{+2}, and so on, can drastically change T_1 values of a sample.

One example of the sort of information that is available from dipolar relaxation studies is given by the study of internal rotations and phase transitions in the alkali metal borohydrides (40). The salient data for $NaBH_4$ are given in Figure 4.4. Both the proton and boron relaxation time measurements clearly show a phase transition at about

190° K. Differentiation of an equation quite similar to equation 4.15 predicts a value of a T_1 minimum in either the boron or the proton curve at a point where $\omega_o \approx 1/\tau_c$.

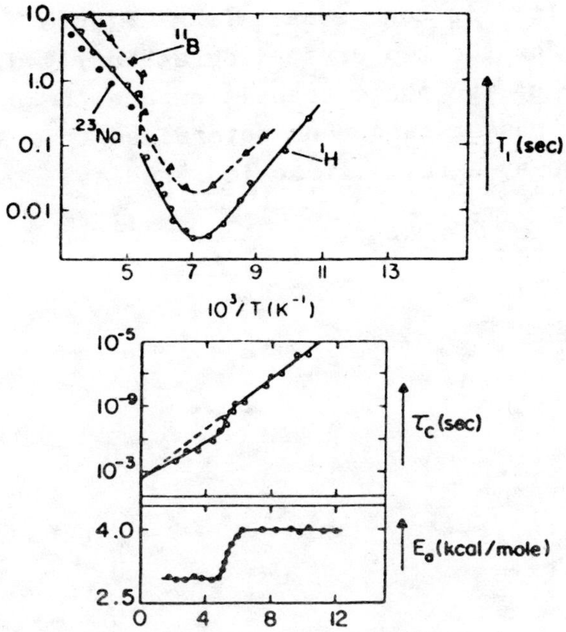

Figure 4.4 (a) Spin-lattice relaxation times T_1 (sec) in $NaBH_4$ vs. temperature ($^{o}K^{-1}$), on a semilog plot. Proton data, O; ^{11}B data, Δ; and ^{23}Na data, ●. (b) The upper trace shows the correlation time τ_c (sec) of the ions as a function of temperature. The lower trace shows the activation energy E_a (kcal $mole^{-1}$) as a function of temperature.

The agreement between experiment and theory is excellent. Knowing the value of τ_c and T_1 in the boron plot at the T_1 minimum also enables one in this case to obtain a value of the boron-hydrogen bond distance. The result, r_{BH} = 1.26 angstroms, is in good agreement with results obtained by other methods (in this case with neutron diffraction measurements of a single crystal sample). A plot of τ_c versus the absolute temperature is shown in Figure 4.4b. The temperature dependence of τ_c may be expressed with good accuracy in the usual form:

$$\tau_c = \tau_o \exp(E_a/RT), \qquad (4.21)$$

where E_a is the activation energy for reorientation of the borohydride anion in the solid state. As can be seen from Figure 4.4b, the value of E_a is quite different for the two crystallographically different forms of the solid. The nature of the phase transition and the crystal structure of the two different phases cannot be determined from NMR measurements, but rather must be done by X-ray diffraction studies.

4.4 Quadrupole Relaxation

Here the Hamiltonian, $\mathbf{H}_{eq}=I*Q*I$, represents the interaction between the nuclear spin (I>1/2) and the electric field gradient at the nucleus. As the molecule reorients, the components Q_{ij} of the quadrupole coupling tensor become random functions of time and provide a relaxation mechanism for quadrupolar nuclei (I>1/2).

For molecular reorientation in the limit that $\omega_0\tau_c \ll 1$, it may be shown that

$$R_1 = R_2 = \frac{3}{40}\left[\frac{2I + 3}{I^2(2I-1)}\right]\left[1 + \frac{\eta^2}{3}\right] \chi^2 \tau_c \qquad (4.22)$$

where η is the asymmetry parameter and $(e^2Qq/h)=\chi$ is the quadrupole coupling constant. As can be seen from equation 4.22, the value of the relaxation time depends upon τ_c and upon the quadrupole coupling constant (e^2Qq/h). Since in mobile liquids the molecular correlation times are 10^{-11} to 10^{-12} sec, the relaxation rate is governed primarily by the magnitude of (e^2Qq/h). In nitrogen compounds, where (e^2Qq/h) for ^{14}N varies from near 0 Hz for the NH_4^+ cation and other species of tetrahedral symmetry to about 4 MHz for CH_3CN, we might expect, then, that the quadrupole contribution to T_1 would vary from hundreds of seconds in NH_4^+ to a few milliseconds in CH_3CN. For CH_3CN, $T_1(^{14}N)$ = 22 msec (41), while for NH_4^+ in aqueous solution, $T_1(^{14}N)$ > 50 sec and is controlled primarily by the N-H dipole-dipole interaction and spin rotation.

In boron compounds a similar situation obtains. For ^{11}B in symmetric molecules, such as $^{11}BH_4^-$, where (e^2Qq/h) for boron is zero (40), T_1 (^{11}B) = 10 sec; in three-coordinate boron compounds, where (e^2Qq/h) = 3 MHz, $T_1(^{11}B)$ values are typically about 10 msec. In ions with tetrahedral or higher symmetry, chlorine or bromine relaxation times can be as long as many seconds. In asymmetric environments these nuclei can have T_1 values of a few microseconds or less. This is due to the very large values (>100 MHz) of the quadrupole coupling constants which are possible for these nuclei.

Since the quadrupole interaction is usually the dominant one for nuclei with spin I>1/2 (unless due to molecular symmetry, e^2Qq/h = 0), and since it is entirely an intramolecular interaction, a measurement of the relaxation time provides an excellent means of measuring simply and unambiguously the molecular correlation time, if the quadrupole coupling constant can be determined independently.

4.5 Relaxation via Chemical Shift Anisotropy

As is well known, a nucleus experiences not the magnetic field B_o, which is generated by the laboratory electromagnet, but rather a field B_{loc} which is given by

$$B_{loc} = B_o - \sigma B_o = B_o(1 - \sigma), \qquad (4.23)$$

where σ is called the shielding factor. This interaction is represented by the Hamiltonian $\mathbf{H}_{cs} = -\gamma h \mathbf{B}^*\Sigma^*\mathbf{I}$, where Σ is the chemical shift tensor. Thus the magnitude of shielding (or chemical shift) at a nucleus is dependent upon the orientation of the molecule in the magnetic field. The rapid molecular motions of molecules in the liquid state average these values out, and one is then usually only concerned with this average value σ where

$$\sigma = 1/3(\Sigma_{xx} + \Sigma_{yy} + \Sigma_{zz}). \qquad (4.24)$$

Although the nucleus on the average sees a chemical shift value represented by σ, on a shorter time scale it sees fluctuations in the local magnetic field. Thus, if Σ_{xx}, Σ_{yy}, and Σ_{zz} are not equal, that is, if the chemical shift tensor is anisotropic, this also provides a relaxation mechanism. Since the magnitude of the chemical shift interaction is field dependent, so is the magnitude of this relaxation mechanism. For a nucleus with axial symmetry it may be shown that

$$R_1 = 1/15\gamma^2 B_o^2(\sigma_\parallel - \sigma_\perp)^2 \; \frac{2\tau_c}{1 + \omega^2\tau_c^2}, \qquad (4.25)$$

$$R_2 = 1/90\gamma^2 B_o^2(\sigma_\parallel - \sigma_\perp)^2 \left[\frac{6\tau_c}{1 + \omega^2\tau_c^2} + 8\tau_c\right], \qquad (4.26)$$

where $\sigma_\parallel$ and $\sigma_\perp$ refer to the shielding along and perpendicular to the symmetry axis. In the absence of axial symmetry the equations become much more complicated. Details are given by Abragam (10).

In the extreme narrowing limit, which is the usual situation in non-viscous liquids, one obtains

$$R_1 = 2/15\gamma^2 B_0^2(\sigma_\parallel - \sigma_\perp)^2\tau_c \ , \qquad (4.27)$$

$$R_2 = 7/45\gamma^2 B_0^2(\sigma_\parallel - \sigma_\perp)^2\tau_c \ . \qquad (4.28)$$

It is of interest that even in the extreme narrowing limit, R_1 and R_2 are not equal, but rather $R_1/R_2 = 6/7$.

Until quite recently, there were few reliable examples of relaxation via chemical shift anisotropy (CSA). Early work on CS_2 produced results which indicated that $T_1(^{13}C)$ was in the range 20-60 sec, a value which clearly could not be explained by intermolecular dipolar relaxation (^{33}S is only 0.75% abundant and has a very small magnetic moment). Since the molecule deviates greatly from spherical symmetry, a plausible explanation of the unexpectedly short value of T_1 was that it was caused by a large chemical shift anisotropy. Other work (42) has shown that the primary relaxation mechanism at 15 MHz is the spin-rotation interaction, which is discussed in Section 4.6. Fifteen years ago, one of the very few experimentally documented cases for relaxation via chemical shift anisotropy was that for $CH_3{}^{13}COOH$, where a plot of R_1 vs. ω_0^2 gave a straight line over the range 9-60 MHz. $T_1(^{13}C)$ drops from a value of 40 sec at 9 MHz (42) to 21 sec at 63 MHz. In general the effect manifests itself only at relatively high magnetic fields, of 2.5 tesla and higher. At fields in excess of 7.0 tesla this interaction can be comparable to the dipole-dipole interaction. Today, (1986) with fields of 10 to 12 tesla becoming relatively common, CSA has become a rather common relaxation mechanism. When this happens, interference effects can arise between CSA and dipolar relaxation. These interference effects can cause the relaxation to become quite complex, and the high-resolution spectra can then contain differentially broadened lines. See references 39a and 39b for further details.

4.6 Scalar Relaxation

As we observed in Section 4.1, any mechanism which gives rise to a fluctuating magnetic field at a nucleus is a possible candidate for a relaxation mechanism. In the case of spin-spin (scalar) coupling between spins I and S (where I = 1/2 and S $\geqslant$ 1/2), the interaction Hamiltonian has the form $\mathcal{H}_S = hI*A*S$. If we consider the relaxation of nucleus I, we see that if B_{nuc} fluctuates in time it can provide a relaxation mechanism. Now B_{nuc} can fluctuate in time in one of two ways: (1) S is time dependent, or (2) A is time dependent.

If nucleus S has a relaxation time which is quite short compared with that provided via the $\mathcal{H}_S$ interaction, that is, if $T_1(S)$ is short compared with 1/A, then the local field $AS(t)/\gamma_I$ produced at nucleus I by a nucleus S fluctuates with a correlation time $\tau_S = T_1(S)$. In this event, of course, only the average value of the spin-coupling interaction is seen, and nucleus I does not exhibit the expected multiplet, but a single line. In a manner similar to that shown above for the dipole-dipole interaction, it may then be shown (10) that

$$R_1(I) = \frac{2A^2}{3} S(S+1) \frac{\tau_S}{1 + (\omega_I - \omega_S)^2 \tau_S^2} \tag{4.29}$$

and

$$R_2(I) = \frac{A^2}{3} S(S+1) \left[\tau_S + \frac{\tau_S}{1 + (\omega_I - \omega_S)^2 \tau_S^2} \right], \tag{4.30}$$

where A is the spin-spin coupling constant expressed in units of radians per second and τ_S is the relaxation time of nucleus S. An excellent example of this situation is provided in the case S > 1/2; S is consequently relaxed primarily by the quadrupole interaction. In this case, which often is referred to as scalar relaxation of the second kind, $T_1(S) = T_2(S) = \tau_S$.

An application of the information available from experiments involving scalar relaxation is given by the measurement of J_{H-Cl} and J_{C-Cl} in the molecule $HCCl_3$ 43). The total relaxation rates for spin I (in this case the proton) are given by

$$\begin{aligned} R_1(I)_{total} &= R_1(I)(DD) + R_1(I)(SC) , \\ R_2(I)_{total} &= R_2(I)(DD) + R_2(I)(SC) , \end{aligned} \tag{4.31}$$

where (DD) refers to the dipolar contribution and (SC) to the scalar coupling contribution. It is found experimentally (44) that $T_1(H) = 90$ sec, T_2 (H) =10 sec, T_1 (^{13}C) =33 sec, and T_2 (^{13}C) = 0.35 sec.

Since τ_S is 34 μsec and $(\omega_I - \omega_S) \simeq 10^8$ rad sec^{-1}, $(\omega_I - \omega_S)\tau_S \gg 1$; consequently, $[1 + (\omega_I - \omega_S)^2\tau_S^2]^{-1} \ll 1$. We can then write

$$R_1(I)\ (\text{total}) = R_1^H(DD) = R_2^H(DD)$$

and (4.32)

$$R_2^H - R_1^H = 1/3\ A^2S(S+1)\tau_S\ .$$

With the value of the chlorine relaxation time T_1 (^{35}Cl) = 34 μsec, we can compute the value for the H-Cl coupling constant from equation 4.32. The result is that $J_{H-Cl} = A/2\pi \simeq 6.9$ Hz. In a similar way, J_{C-Cl} is found to be 23 Hz.

This same sort of information is also available via rotating reference frame ($T_{1\rho}$) experiments, which are discussed later.

Scalar relaxation can, in general, occur in a two-spin (I,S) system whenever I and S are coupled. In the case (^{13}C) > T_1 (H), when R_1 (H) = $1/T_1$ (H) $\ll 2\pi J = A$, one observes that T_2 (^{13}C) < T_1 (^{13}C) for the ^{13}C nucleus in question. For example, in $^{13}CH_3I$ where J_{CH} = 152 Hz, T_1 (^{13}C) = 13.4 sec, T_2 (^{13}C) = 3.9 sec, and T_1 (H) = 10.9 sec; while in $^{13}CH_3COOCD_3$, J_{C-H} = 130 Hz, T_1 (^{13}C) = 19.2 sec, T_2 (^{13}C) = 6.1 sec, and T_1 (H) = 12.5 sec.(45).

Scalar relaxation plays a salient role in the broadening of ^1H lines for protons bonded to a quadrupolar nucleus, S, such as ^{14}N and ^{11}B, where τ_S = $T_1(S)$ lies in the proper range (10-100 msec). Such effects are not observed with nuclei such as chlorine and bromine because these nuclei have, as a rule, much shorter relaxation times. A simple calculation using equation 4.33 shows that for reasonable values of the spin coupling (100 Hz), when $T_1(S)$ (quadrupole) $\leqslant 10^{-5}$ sec, the effect of scalar relaxation on the proton line width [i.e.,$R_2(I)$] is negligible.

Scalar relaxation can also occur when A becomes a function of time. This situation is often referred to as scalar relaxation of the first kind and arises when chemical exchange is present. In this case the local field at I is AS/γ_I when I and S are covalently bound in the same molecule, and zero otherwise. If the chemical exchange rate is much larger than either the coupling, A, or $1/T_1$ for either I or S, and if the time the nuclei are uncoupled is short compared with the time they are coupled, then the multiplet structure disappears and only a single resonance line is observed. This is analogous to scalar relaxation of the second kind discussed above, and in fact equations 4.29 and 4.30 are valid here too; τ_S now represents τ_e, the exchange time, not the relaxation time $T_1(S)$.

If τ_S or τ_e is not short enough to cause the multiplet to collapse completely to a single narrow line (a few hertz or less), a more detailed analysis is needed to obtain useful chemical information (46, 47).

The fact that these equations are identical for the case where one replaces τ_e by τ_S is quite reasonable, for all nucleus I experiences is a local field of magnitude AS/γ_I which is fluctuating in a characteristic time τ. It is immaterial to nucleus I whether this fluctuation arises from a chemical exchange process or from a fast relaxation process for nucleus S.

One might well ask whether this scalar interaction should be expected to be a very effective relaxation process, since the scalar coupling interaction is usually small ($H \approx 10^4$ Hz) compared with the dipole-dipole interaction ($H \approx 10^8$ Hz). The answer is that it can be very effective under the proper conditions. Although H_d is generally much larger than H_S, the molecular correlation time τ_c is, for most liquids, rather small ($\tau_c \approx 10^{-11}$ sec). On the other hand, τ_S and τ_c can be, and often are, in the range where $(\omega_I - \omega_S)\tau \approx 1$. In this case the scalar interaction is greatly enhanced and is sometimes the dominant relaxation mechanism. It is relatively easy to determine whether or not this is the case, since the relaxation, as can be seen from equations 4.29-4.32, is then frequency dependent.

For completeness we should mention that, in principle, if A is anisotropic, the anisotropy of the spin-coupling tensor together with molecular motion also produces fluctuating fields at the coupled nuclei. However, this anisotropy is almost always much smaller than the dipole-dipole interaction, and since the spin-coupling and dipolar correlation times are comparable, the contribution of the spin-coupling anisotropy to R_1 is negligible.

4.7 Spin-Rotation Relaxation

The spin-rotation interaction arises from magnetic fields generated at a nucleus by the motion of a molecular magnetic moment which arises from the electron distribution in a molecule. Let us consider one particular electron and a given nucleus. As the molecule rotates about the nucleus at a radius, say, R. The rotational frequency V is given by

$$V = hJ/2\pi \ , \qquad (4.33)$$

where the molecule is in the Jth rotational state and I is the moment of inertia of the molecule. The current generated by the electron is then

$$i = (e/c)V \ , \qquad (4.34)$$

and the magnetic moment associated with this circulating current is

$$\mu_J = i(\pi^2) = (eh/2\pi M_c)J \cong \mu_N J \qquad (4.35)$$

and $I = MR^2$, where M is the nuclear mass and μ_N is the nuclear magneton. This magnetic moment generated by the motion of the electron produces a local magnetic field at the resonant nucleus of the order of $\mu_N J/R^3$. For the hydrogen molecule this local field is about 35 gauss and for HCl the field at the hydrogen nucleus is about 10 gauss. Since this effect is proportional to the rotational velocity, and inversely proportional to the moment of inertia of the molecule, we may anticipate that in general the smaller the molecule, the more important will be the spin-rotation interaction. Furthermore, symmetric molecules with little or no intermolecular interactions (such as hydrogen bonding) will be most affected, since they will have somewhat greater angular velocities.

The Hamiltonian for this interaction, $\mathbf{H}_{sr}$, is given by

$$\mathbf{H}_{sr} = -h\mathbf{I}*\mathbf{C}*\mathbf{J} \ , \qquad (4.36)$$

where I is the nuclear spin operator, J is the molecular rotation, or angular momentum operator, and C is the spin-rotation tensor. In this case, collisions cause the molecule to experience changes in the direction (M diffusion) or in both the direction and the magnitude (J diffusion) of the angular momentum. This in turn causes fluctuations in the local field at the resonant nuclei in the molecule. For liquids undergoing isotropic molecular reorientation the relaxation rate is given by

$$R_1 = (2\pi IkT/h^2)C_{eff}^2\tau_J, \qquad (4.37)$$

where $C_{eff}^2 = 1/3(2C_\perp^2 + C_\parallel^2)$ and τ_J is the angular momentum correlation time, which is a measure of the time a molecule spends in any given angular momentum state ($C_\parallel$ and $C_\perp$ are the principal components of the spin-rotation tensor parallel and perpendicular to the symmetry axis, respectively).

It may be shown (48, 49) that the angular momentum correlation time, τ_J, is related to the molecular reorientation correlation time, τ_c, which was discussed in the previous cases, by the equation

$$\tau_c \tau_J = I/6kT , \qquad (4.38)$$

where I is the moment of inertia of the molecule, k is the Boltzmann constant, and T is the absolute temperature. It should be noted that equation 4.38 only holds at temperatures well below the normal boiling point of a liquid.

The important distinction between this relaxation mechanism and the others discussed is that τ_J becomes longer as the sample temperature increases, whereas τ_c becomes shorter. As the temperature becomes quite high and the sample becomes a gas, collisions become more infrequent and the molecule remains in a given angular momentum state for a longer time. On the other hand, the higher the temperature, the faster a molecule reorients and the shorter becomes τ_c. The result of this is, of course, that the relaxation time T_1 for the spin-rotation interaction becomes longer as the temperature decreases. This behavior is opposite that observed for the other relaxation mechanisms.

An example of a system in which spin-rotation is an important relaxation mechanism is given in the ^{35}Cl and ^{19}F pulse NMR relaxation study of neat ClO_3F (50). In this system the fluorine spin-rotation interaction eclipses all other ^{19}F relaxation mechanisms in its importance, and the chlorine nucleus is relaxed almost entirely by the quadrupole interaction. Since the chlorine quadrupole coupling constant and the effective fluorine spin-rotation constant are known from independent work, one can accurately measure both the molecular reorientation correlation time τ_c and the angular momentum correlation time τ_J.

In general, when one is dealing with small, symmetric molecules and with nuclei which have a large range of chemical shifts (e.g., ^{19}F, ^{13}C, ^{15}N) one can anticipate that spin-rotation interactions might well be important. The apparent relation between R_1 (spin-rotation) and the chemical shift arises because both the chemical shift and the spin-rotation tensor components of any given molecule depend on the electron distribution in a molecule. A distribution which results in large chemical shifts will also lead to large spin-rotation interactions. Methyl group carbon nuclei normally have a large contribution to their relaxation from the spin-rotation interaction.

4.8 Summary of Relaxation Mechanisms

In Table 4.1 we have summarized the various relaxation interactions and given some approximate values for the interaction energies and relaxation times. We have also listed some examples of molecular systems, with the appropriate nucleus in boldface type, for which the interaction in question is important.

Table 4.1

Summary of Relaxation Mechanisms

Interaction	T_1 Range[a]	E_c[b]	Typical Systems
Dipole-Dipole	$1 \leq T_1 \leq 100$ sec	$\gamma_I \gamma_S \hbar / r^3$	$PHO_3^=$, CH_3I $HMn(CO)_5$
Quadrupole	$10^{-7} \leq T_1 \leq 10^2$ sec	$(e^2Qq/\hbar)$	NH_3, Na^+, D_2O
Chemical Shift	$1 \leq T_1 \leq 100$ sec	$\gamma B_o(\sigma_\parallel - \sigma_\perp)$	$PFO_3^=$, $H^{13}C{\equiv}CHC$
Scalar Coupling	$0.1 \leq T_1 \leq 100$ sec	$\frac{2A^2}{3}[S(S+1)]^{1/2}$	$^{13}CHBr_3$, PBr_3
Spin Rotation	$10^2 \leq T_1 \leq 100$ sec	$(C_\parallel - C_\perp)/h$	ClO_3F, $^{15}NH_4^+$

[a] For $\tau_c \simeq 10^{-11}$ sec. [b] $1/T_1 = R_1 = E_c^2 \tau_c$.

For a discussion of the relation between the chemical shift tensor and the spin-rotation tensor see Maryott et al. [50] and Deverell [51].

4.9 Some Practical Considerations

Equation 4.15 can be very useful, since it gives one a method for measuring internuclear distances and molecular correlation times of molecules in solution. Consider, for example, a sample of $HMn(CO)_5$ in d8-toluene solution [50b]. This molecule is an example of a simple two spin system. We represent the proton spin by I (I = 1/2) and the manganese spin by S (S = 5/2). In this case, the only molecular motions that contribute to the relaxation are the motions about axes perpendicular to the C_{4v} symmetry axis (only motions about these axes cause the angle between the H-Mn internuclear vector and the magnetic field to change). Consequently, there is only a single correlation time for relaxation. For the calculation of internuclear distances or the calculation of correlation times, equation 4.15 is not in a very convenient form. A more useful form of equation 4.15 is given by

$$R_1(I) = (\mu_o^2/160\pi^2)\,\gamma_I^2\gamma_S^2\,\hbar^2 S(S+1)r^{-6}\ \times$$

$$\left[\frac{\tau_c}{1+\omega_-^2\tau_c^2} + \frac{3\,\tau_c}{1+\omega^2\tau_c^2} + \frac{6\,\tau_c}{1+\omega_+^2\tau_c^2}\right] \qquad (4.39)$$

As can be seen from equation 4.39, for any given experiment there are only four variables: the relaxation rate $[R_1(H)]$, the proton-manganese internuclear distance (r), the correlation time (τ_c), and the magnetic field strength (B_o) [B_o determines both $\omega(H)$ and $\omega(Mn)$]. To proceed further, we must consider the units used with equation 4.39. In all of the relaxation calculations in this book we use SI units. In SI units the required constants are:

μ_o = $4\pi \times 10^{-7}$ m kgm sec^{-2} A^{-2}

h = 6.62517×10^{-34} m^2 kgm sec^{-1}

γ_H = 2.67530×10^8 kgm^{-1} sec A

γ_{Mn} = 0.65982×10^8 kgm^{-1} sec A

the internuclear distance r is measured in meters, the correlation time

τ_c in seconds and the relaxation rate R in sec^{-1}.

At room temperature τ_c in a typical nonviscous liquid has a value of about 1 ps (picosecond). Cooling the sample and slowing down the molecular motions can cause τ_c to become relatively long, a few hundred microseconds or longer. Simple arithmetic shows that at 500 MHz $R_1(H)$, is a maximum when $\omega(H)\tau_c = 0.924$. At this R_1-maximum point, for a given field strength, there are only two variables, $R_1(H)$ and r. A measurement of the proton relaxation rate, then, should give an accurate value for the bond distance. Since $R_1(H)$ varies as the inverse sixth power of the bond distance, an error of six percent in $R_1(H)$ results in a one percent error in the bond distance r. In actual practice, the quite reasonable assumption is made that the correlation time is related to the temperature by the equation

$$\tau_c = \tau_o \exp(-E_a/RT) \tag{4.40}$$

where τ_o is a constant, E_a is the activation energy for molecular reorientation, T is the temperature in degrees Kelvin and R is the gas constant. A least squares fit of the measured R_1 values as a function of the temperature will then provide the best value for r, E_a and τ_o. For $HMn(CO)_5$ a value of r = 0.165 ± 0.005 nm was obtained. This value for r agrees well with the value of 0.160 nm obtained from neutron diffraction studies.

CHAPTER FIVE

FOURIER TRANSFORM CALCULATIONS

In Chapter 1 we gave a few brief examples of the response of the spin system to a brief, intense burst of r.f energy whose frequency was equal to the Larmor precessional frequency of the nuclei in the sample. We observed that the free induction decay (FID) signal following the r.f. pulse contains all of the information available about the number of nuclei present, the relative intensities and their line widths. Because this FID, or time response, contains the information in a form that is not very useful, spectroscopists have found it necessary to calculate the Fourier transform of the time response. In this chapter we consider Fourier transform methods in more detail and look into some of the practical aspects of this form of spectroscopy and some of the limitations which the mathematical formalism imposes upon the data.

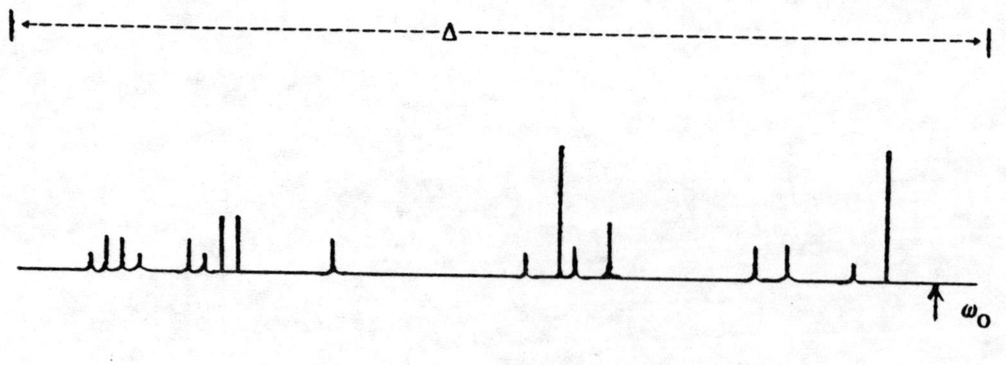

Figure 5.1 Schematic representation of an NMR spectrum showing the range, Δ, of frequencies occurring in the spectrum. If an r.f. pulse is applied at frequency ω_0, it will excite all frequencies in the range $\omega_0 \pm (1/t_p)$. The pulse could also be applied at the center of the spectrum.

In general, one would like to record the spectrum of a sample which exhibits a number of spectral lines at different frequencies and with different intensities and line widths. A "typical" spectrum is shown in Figure 5.1. At a magnetic field of B_o (tesla) this spectrum covers a range of Δ hertz. In the pulse NMR experiment one does not sweep the frequency of the irradiation field to observe which spectral lines are present. Rather, one uses a source with fixed frequency ω_o which lies in the proximity of the spectral region (see Figure 5.1). To see how the r.f. field of a pulse can interact with nuclear moments which are precessing at a frequency ω, which is quite different from ω_o, let us examine the behavior of the magnetization in a frame rotating at the frequency ω. In the rotating reference frame, the magnetization **M**, which arises from nuclei precessing about the static field at angular frequency ω_i, precesses about the effective field, which as given in equation 1.59 is

$$|B_{eff}| = (1/\gamma)[(\omega_i - \omega)^2 + (\gamma B_1)^2]^{1/2} . \qquad (5.1)$$

If B_1 is chosen large enough so that

$$\gamma B_1 >> 2\pi\Delta, \qquad (5.2)$$

where Δ, in hertz, spans the entire range of chemical shifts of the nuclear species present in the sample, then at any ω_i within the spectrum the term $(\omega_i - \omega)$ can be neglected; we may then write

$$B_{eff} \approx B_1 . \qquad (5.3)$$

In other words, the magnetization **M** precesses about B_1, which is usually taken to be directed along the x' axis. This is true for all of the nuclei with Larmor precessional frequencies in the range Δ. For a $\pi/2$ (or 90°) pulse, we know from equation 2.1 that

$$\gamma B_1 t_{\pi/2} = \pi/2 \qquad (5.4)$$

By making use of the condition given in equation 5.2 above, we see that $t_{\pi/2}$ must be quite short, namely,

$$t_{\pi/2} << 1/(4\Delta) \text{ sec.} \qquad (5.5)$$

A more descriptive way to view this situation is to consider the pulse envelope as a square wave. Then, according to equation 1.44, it contains Fourier components covering the approximate range of frequencies $v_o \pm 1/t_p$, where $2\pi v_o = \omega_o$ is the applied r.f. field expressed in units of of Hertz (ω is in units of radians). Here again, one sees that a short r.f. pulse is equivalent to the simultaneous application of a wide range of r.f. voltages at radio frequencies, covering the range $\pm(1/t_p)$. It should, perhaps, be pointed out that the frequency range spanned by the pulse is dependent <u>only</u> on the duration of the pulse; it is quite independent of the r.f. power. If the r.f. power is very small, the magnetization **M** is not tipped very far from its equilibrium state in 1 μsec. If t_p = 1 μsec, however, the range of frequencies spanned is about ± 250,000 Hz. But note that this condition alone says nothing about the magnitude of γB_1. It is the 90° pulse time that is related to γB_1. If $t_{\pi/2}$ is 1 μsec, then $\gamma B_1 \approx 60$ gauss $\approx 250{,}000$ (for protons).

5.1 Sensitivity Advantages of Pulse NMR Methods

Ernst and Anderson (7) have made a careful study of the conditions under which Fourier transformation of the FID signal can provide spectral information more efficiently manner than direct frequency-sweep experiments of that same spectrum. They showed that for equivalent signal/noise ratio (S/N), the Fourier transformation provides a saving in time compared with slow passage frequency (or field) sweep methods of approximately Δ (we assume that the resolution, r, is the same for both methods), where Δ is the range of frequencies to be observed and $r = v_{1/2}$ is the width of a typical line in the spectrum. This is certainly a reasonable expectation if we consider that the time to acquire a cw frequency sweep spectrum is $t_{cw} \approx \Delta/r$ and the time to acquire an FID signal with resolution r is $t_{ft} \approx 1/r$. The time advantage factor, ε_t, of the Fourier transformation over the cw methods is then just the ratio of t_{cw} to t_{ft}, or

$$\varepsilon_t = t_{cw}/t_{ft} \, , \qquad (5.6)$$

$$\varepsilon_t = \Delta \, .$$

The sensitivity advantage factor, ε_s, is simply the square root of the time advantage factor

$$\varepsilon_s = (\varepsilon_t)^{1/2} \, . \qquad (5.7)$$

Here, then, lies the great power of the pulse NMR method. The time required to obtain a spectrum is independent of the width of the spectrum, but depends only on the spectral resolution which is required.

For example, suppose that we wish to acquire a carbon-13 spectrum at a frequency of 50 MHz (B_0 = 4.7 Tesla). For typical carbon-13 spectra the spectral range Δ is about 200 parts per million (ppm). At 50 MHz, then, Δ is about 10,000 Hertz. Suppose that we desire a spectral resolution of 1 Hz. The time required to obtain the cw spectrum is then Δ/r = 10,000/r = 10,000 sec. ≈ 2.5 hr. The time required to obtain the FT spectrum is 1/r = 1 sec. It is clear that without the development of pulse and FT-NMR methods, carbon-13 NMR spectroscopy would not be the routine procedure that it is today.

Since most of the slow passage carbon-13 spectra taken in the past were taken at very rapid sweep rates, the time advantage factors realized were not as great as they were predicted to be. With the rapid sweep methods, however, there was considerable distortion of the spectrum. By contrast, the pulse NMR methods yield a big time advantage (perhaps a 1000 fold advantage for the carbon-13 example cited above) and, importantly, they give spectra of much higher quality. In addition, in the past few years a number of sensitivity enhancement experiments have been developed which give an additional sensitivity advantage to the pulsed NMR experiments, which are not attainable using cw methods. These sensitivity enhancement experiments are discussed in detail in Chapter 6.

Some of the more important advantages of pulse NMR spectroscopy are:

1) The signal/noise (S/N) ratio can be improved by repeating the experiment a number of times and coherently adding successive FID signals before calculating the Fourier transform.

2) The spectra of transient or moderately short-lived species can be observed.

3) Optimum homogeneity of the magnetic field can be more easily maintained for the relatively short pulse experiment than for the very much longer cw experiments.

4) Selective sensitivity enhancement experiments can be easily executed with pulse NMR methods. Such experiments are impossible with cw methods.

5) A wide variety of relaxation time studies can be done using pulsed methods. These experiments provide a wealth of information about molecular structure, electronic structure and molecular dynamics.

6) Two-dimensional FT-NMR experiments provide a wealth of information about complex systems. Useful NMR data of such systems was totally out of the question before the development of these 2D-FT-NMR methods.

5.2 Signal Processing

The computer plays a central role in the modern pulse NMR spectrometer. In addition to the control functions discussed in Chapter 3, the computer must record data and carry out a wide variety of data analysis tasks. The first step is, of course, to record the data.

Data are acquired through the use of an analog-to-digital converter (ADC). The ADC simply converts the analog signal present in the receiver into a number of digital data points. This is done in exactly the same way that one records signals with the now common digital voltmeter. However, the rate at which one makes the receiver voltage readings can sometimes be quite high (up to 200,000 measurements per second). These measurements are made at time intervals that are accurately controlled by a special digital clock. A typical ADC has a 12 bit dynamic range and, consequently, has a decimal dynamic range of 4096 with a resolution of unity. The acquisition of the data must be at precisely known intervals of time and, for FT spectroscopy, they must be precisely equally spaced. Any uncertainty or "jitter" in the timing will cause substantial spurious, unwanted signals in the transformed frequency domain signal. For a complete discussion of such details, see reference 52.

Probably the most important parameter in characterizing the receiver response is the sampling rate. The Nyquist theorem (53) states that the highest frequency in a given time response must be sampled at a rate of at least two points per cycle in order to be represented accurately. If a spectrum contains frequencies from 0 Hz up to Δ Hz (see Figure 5.1), then one must sample at a rate of 2Δ samples per second. The highest frequency which can be accurately represented by sampling at a rate of 2Δ

Hz per second is thus Δ Hz. This frequency, Δ, is known as the Nyquist frequency.

Figure 5.2a shows a sine wave of frequency Δ Hz sampled at the Nyquist frequency. Figure 5.2b shows a sine wave of somewhat higher frequency, $(\Delta + f)$ Hz. Figure 5.2c shows a sine wave of somewhat lower frequency, $(\Delta - f)$ Hz. Note that the dots, which represent the sampling times, indicate what is "observed" by the computer. Thus the computer "sees" the same result for both of the frequencies $(\Delta + f)$ and $(\Delta - f)$, and it also sees the same amplitude for both waveforms. This is the result of not observing the Nyquist requirement. This phenomenon is sometimes referred to as foldback or aliasing. It should be noted that folded or aliased signals also usually experience a phase shift which is related to the magnitude of the foldback.

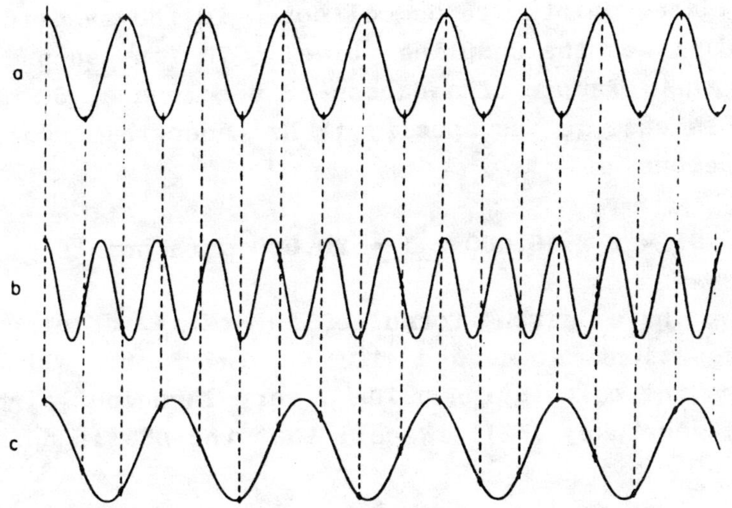

Figure 5.2 (a) Sampling is done at twice the Nyquist frequency, 2Δ, where Δ, the Nyquist frequency, is the value of the highest frequency component of the waveform sampled. (b) If samples are taken at a rate slightly less than Δ then the sampled waveform appears as a lower frequency (c).

The first condition to meet is that the sampling rate, SR be equal to or greater than 2Δ, where Δ is the **maximum** frequency to be sampled. The second condition relates to the sampling time or the acquisition time. If one requires a resolution (line width) of r Hz, then one **must** sample for 1/r sec . This condition is plausible if one considers that in order to distinguish between 1000 Hz and 1001 Hz with a frequency counter, one must use a gate time (sampling period) of at least 1 sec. With a frequency counter, one is counting the number of zero crossings of the two wave forms in a given time interval, the gate time. One does a similar measurement using FT methods. One of the implicit (but never openly admitted) conditions of the Fourier transform method is that all frequencies contained in the time response must be harmonics of 1/ST, where ST is the sampling time (we consider this point in more detail later).

The frequency range, Δ, and the resolution, r = 1/ST, together set the requirement for the size, SI, of the data array, since

$$SI = (2\Delta)(ST). \tag{5.8}$$

The number of data points to be collected is the sampling rate (data points per second) times the sampling time, ST (in seconds). Let us consider again the example of a carbon-13 spectrum at 50 MHz where $\Delta \simeq$ 10,000 Hz; in this case let us specify 1 Hz spectral resolution. The size, SI, is therefore

$$SI = 2 \times 10{,}000 \times 1 = 20{,}000 \text{ data points}$$

Thus far we have given formulae in terms of discrete Fourier transforms (see equations 1.44 and 1.45). In fact, all current FT spectroscopy does the calculation using a very ingenious algorithm first devised by Cooley and Tukey (52). Recall that the DFT is defined by

$$I_f = \sum_{t=0}^{N-1} G_t \exp(-2\pi i \omega_f t/N), \quad f = 0,1,2,3,\ldots,N-1 \tag{5.9}$$

where I_f is the amplitude or intensity at frequency ω_f in the frequency domain, G_t is the amplitude at time t in the time domain, and N is the total number of data points. From elementary trigonometry

$$e^{-i\phi} = \cos\phi - i\sin\phi \tag{5.10}$$

Hence equation 5.9 can be written as the complex sum of sine and cosine transforms. For a discrete Fourier transform (DFT) any arbitrary number of data points may be used.

The Cooley-Tukey algorithm, usually referred to as a fast Fourier transform (FFT), is a much more efficient way of calculating the transform. The FFT method involves repetitive pairwise sorting of the data points and requires $2N\log_2 N$ multiplications for N data points. In contrast, the DFT requires N^2 multiplications (see equation 5.9). Because of the pairwise sorting procedure used in an FFT calculation, it is necessary to use 2^n data points, where n is an integer. For 4096 (4K) data points the FFT requires about 170 times fewer multiplications than the DFT. Of course, even greater savings are made with larger values of N. A further advantage of the FFT comes from its efficient use of computer memory. Because the coefficients are evaluated by an iterative procedure in which intermediate and final values can be written over the original data, only slightly more than N memory locations are needed in total to carry out an N point transformation (where $N = 2^n$). Since we require that $N = 2^n$, we must use an 8K data array if 4K (K = 1024) data points are not sufficient. In the carbon-13 example given above we therefore cannot use 20,000 data points, but instead we must choose 16K or 32K data points. Consequently, we must use a slightly shorter sampling time or a much longer sampling time. For more details on FFT calculations, see reference 52.

Although the FFT is one of the major computational tasks of the computer, there are a number of other important ones. After the data acquisition is finished, the first computational step is usually to calculate the average value of the signal and to subtract this value from all of the data values in order to remove any dc component that contains no spectral information but can contribute a large peak in the transformed spectrum at the ω_0 frequency (zero frequency in the spectrum). This is usually referred to as a baseline correction. Other important steps in the data reduction process are:

1) Exponential multiplication. If the FID is multiplied by a factor of the form exp(-at), where a is a constant and t is time, the FID will experience a decreasing intensity as a function of time. Since the signal is decreasing exponentially and the average noise level is constant throughout the FID, the result will be to improve the S/N in the transformed spectrum. The price one pays for such filtering is some artificial broadening of the spectral lines. A positive exponential filter can also be employed. The result of this is to improve, sometimes dramatically, the resolution in the transformed spectrum. As might be

expected, the price for this improved resolution is a much poorer signal to noise and some distortion of the weaker peaks in the spectrum.

2) Zero filling. Additional zeros may be added to the FID to create a longer apparent sampling time. Since r = 1/ST, zero filling has the effect of improving the resolution without decreasing the S/N. If the actual FID contains N actual data points and we add N additional zeros, we actually increase the resolution by a factor of two. Adding even more zeros will increase the spectral definition but not the real resolution.

3) Phase correction. After the FFT a correction must be made for phase and amplitude distortion introduced by imperfections in the spectrometer hardware. There are three types of phase error commonly introduced into the experiment:

a) A constant error independent of frequency is caused by misadjustment of the phase detector so that a mixture of absorption and dispersion modes is produced, exactly as in a conventional cw experiment. Since the process of the FFT generates both the sine and cosine transforms separately, the pure absorption mode spectrum $A(\omega)$ may be generated from a linear combination of the sine (S) and cosine (C) transforms:

$$A(\omega) = pC(\omega) + (1 - p^2)^{1/2} S(\omega) . \qquad (5.11)$$

We show below how the value of p may be calculated.

b) The finite length of the $\pi/2$ pulse that initiates the FID along with the finite receiver recovery time causes substantial delays in the time between the start of the pulse and the time at which data recording commences. This leads to a phase error that varies linearly with frequency. Correction for this type of error can be made in several ways. One method is to express the coefficient p in equation 5.11 above as a sum of two parameters p_0 and p_1 in the following way:

$$p = p_0 + p_1(\omega - \omega_0) . \qquad (5.12)$$

c) Other amplitude and phase errors result from signal processing in r.f. and i.f. amplifiers, phase detectors, and r.f. and audio frequency (a.f.) filters. All of these electronic artifacts can be corrected for by proper adjustment of the p_0 and p_1 parameters.

The final amplification stage is usually the a.f. amplifier. This is usually followed by a Butterworth filter, which has the characteristic that it passes all frequencies from 0 Hz up to a defined maximum frequency; all higher frequencies are filtered out. This is known as a low-pass filter; that is, the filter has a sharp cutoff frequency ω_{max}. Its phase characteristics are rather complex, but here, too, phase distortions can be corrected by proper adjustment of p_0 and p_1. Whatever means is used to measure the dependence of p on frequency, the numerical values of p_0 and p_1 for any given spectrum can be calculated in a straightforward manner. In most cases an interactive program displays the spectrum for given values of p_0 and p_1 on the spectrometer video display and allows the operator to select the values for p_0 and p_1 (usually via two separate knobs on the console) that will result in the best looking absorption spectrum.

5.3 <u>Other Data Analysis Methods</u>

The Fourier transform is a stable and rapid procedure for converting a time response, or interferogram, into a frequency domain spectrum, and it has played a vital and important role in making modern pulse NMR a powerful and versatile tool for the study of molecular structure and molecular dynamics. Unfortunately, Fourier transformation does have several weaknesses.

Inspection of equation 5.9 indicates that in a Fourier transformation, the time response data is treated as a linear combination of a limited set of discrete sine and cosine functions, and the resolution in the frequency domain is limited by the reciprocal of the sampling time. For this reason, in a Fourier transformation, resolution is improved by a longer sampling time. But, since the signal decays in time while the average noise is more or less constant, better resolution is obtained at the expense of a poorer S/N value. In an attempt to obtain better S/N performance, better resolution, and less distortion of the original time response data, a number of workers have been applying maximum entropy methods (MEM) and auto-regressive methods (ARM) to NMR data.

In most modern pulse NMR spectrometers the data consists of a set of experimental values measured at equal intervals in time. Thus the data

$X_o=\{x_o, x_1, x_2, \ldots, x_M\}$ where the x_n are experimental values recorded at times $t_n=n\Delta\tau$. The goal is to obtain from this data a spectral representation of the original signal present in the frequency range $-f_N \leq f \leq +f_N$, where f_N is the Nyquist frequency $[f_N = 1/(2\Delta\tau)]$. If we represent the spectral function by S(f), then

$$x_n = \int_{-f_N}^{+f_N} S(f)\, \exp(-i\lambda n) df \quad (\lambda=2\pi f\Delta\tau). \qquad (5.13)$$

The Fourier transformation is one way to obtain an estimate of S(f). Implicit in the FFT is the assumption that $x_n=0$ for all data outside the sampling region. This assumption is not valid and leads to distortion of the frequency domain data. The MEM is essentially a modified Fourier transformation subject to the mathematical constraint that our knowledge of time data outside the sampled region is a minimum, or, our lack of knowledge (entropy) outside the sample region is a maximum. At its best, the MEM can, for a given set of experimental data, provide better resolution and S/N in the frequency domain than Fourier transformation. For full details of MEM see references 53a and 53b.

There are, in this author's view, two drawbacks to MEM. First, the basic premise of the MEM is, for an NMR experiment, incorrect, since we do have knowledge of the time response outside the sampled region. We know that the time response is an ensemble of damped cosine functions before, during, and after the sampling period. Second, MEM calculations are not stable. That is, the number of peaks obtained in the frequency spectrum depends on how many coefficients are used for the MEM calculation. In our laboratory, we have obtained from the same set of experimental time domain data as many as eleven lines and as few as four depending upon the order of the MEM calculation. Our experience and that of several other research groups is that in the current state of development (early 1987), MEM calculations are not sufficiently stable for routine, meaningful calculations.

Auto regressive methods (ARM) are, in my view, superior to MEM for two reasons. First, they are stable and give reliable reproducible results and, second, they follow Able's first rule of data analysis: the result of any transformation imposed on the experimental data shall incorporate and be consistent with all relevent data (53a). In the ARM the signal, which is sampled at regular intervals $n\Delta t$, is assumed to consist of a set of K exponentially damped sinusoids plus white noise. The data can therefore be written in the form

$$x_n = x(n\Delta t) = \sum_{k=1}^{K} c_k \exp(-b_k n\delta t)\cos(\omega_k n\Delta t + \phi_k) + w_n. \qquad (5.14)$$

For each frequency, K, there are four degrees of freedom: a frequency ω_k, an amplitude c_k, a line width or damping factor b_k, and a phase ϕ_k. The noise is represented by w_n. Using a linear prediction (LP) model one assumes that each data point can be expressed as a linear combination of the M previous data points. Thus

$$x_n = a_1 x_{n-1} + a_2 x_{n-2} + \cdot \cdot \cdot + a_M x_{n-M} \qquad (5.15)$$

where n can have any value between M and N-1 and the LP coefficients a_m (m=1,2,...,M) are independent of n. For a single, exponentially damped sine wave without noise it is straightforward to show that

$$x_n = \exp(bn\Delta t)\cos(\omega n\Delta t + \phi), \qquad (5.16)$$

and

$$a_1 = 2\exp(-b\Delta t)\cos(\omega\Delta t),$$
$$a_2 = -\exp(-2b\Delta t).$$

a_1 and a_2 are independent of n. It can also be shown that equation 5.15 holds exactly for any arbitrary number, K, of frequencies, if no noise is present. In this event, there are 2K LP coefficients. If the signal contains noise, the number of LP coefficients required is much greater than 2K and sometimes as many as 0.75N coefficients are required (where N is the total number of data points). In practice, the LP coefficients are obtained by solving equation 5.15 in a least squares fashion. The values for the b_k and ω_k are related to the a_M (the LP coefficients) and are obtained by finding the roots of the polonomial equation

$$Z^{2K} - a_1 Z^{2K-1} - \ldots\ldots - a_{2K} = 0 \qquad (5.17)$$

The c_k and ϕ_k parameters are then obtained from a separate set of equations generated from equation 5.14 (equation 5.14 is written down for all N data points). These equations are solved by another least squares procedure for the c_k and ϕ_k. This entire process is quite stable and gives excellent results. For an FID signal with a large number of frequencies present and/or a poor S/N value, the calculation can be very time consuming. For a given number of data points, the LP method gives a spectrum with much better resolution and a better S/N value than that obtainable via Fourier transformation. It is especially useful in situations where the S/N is poor or where good resolution is needed, but only a restricted number of data points is available. The details are explained clearly in an excellent paper by Barkhuijsen el al. (53c and 53d).

CHAPTER SIX

SENSITIVITY ENHANCEMENT TECHNIQUES

6.1 Sensitivity Enhancement Experiments

Although pulse NMR methods and the FFT have made enormous improvements in the sensitivity of NMR experiments, there is still a need for even greater improvements in sensitivity. In our discussion above we assume that the $\pi/2$ pulses can be repeated immediately after each cycle of FID data collection. If the sampling time is primarily limited by field homogeneity, then we might want to limit the sampling time to about $3T_2^*$, where $\pi T_2^* \simeq \Delta B_0$, the magnetic field inhomogeneity. If $T_1 \simeq T_2^*$ then such a rapid pulse repetition rate should be satisfactory. Usually, however, $T_1 >> T_2^*$, and very little restoration of M_z toward its equilibrium value of M_0 occurs. This is especially true for carbon-13, silicon-29, nitrogen-15 and other nuclei with small γ values. Since T_2^* is often limited by magnetic field homogeneity to about 1 sec, while T_1 may be of the order of 10 - 100 sec, this situation is of practical importance.

A number of different methods have been proposed to overcome the undesirable signal attenuation by saturation due to magnet inhomogeneity and long T_1 values. One commonly used method is to add a small amount of paramagnetic ions to the solution. This substantially shortens the T_1 values and allows much more rapid recovery of the M_z magnetization. A second method utilizes the fact that the Fourier transform relationship between the FID and the cw spectrum is true for an arbitrary tip angle, t_p, and not only for $\pi/2$ pulses. If the pulse angle is shortened while γB_1 is held constant, then **M** rotates through a smaller tip angle θ. The result is that the amplitude of the projection of **M** in the xy plane is attenuated by a factor of sin θ. However, M_z is only diminished by the amount $M_0(1 - \cos\theta)$. For small θ, $\sin\theta \ll (1-\cos\theta)$. Consequently, the attenuation of M_{xy} and the resultant loss of signal is more than offset by the smaller time required to reestablish equilibrium along the z axis. The optimum value of θ depends on T_1 and T_2^* but is usually determined empirically as the value that leads to the optimum signal (7). This is sometimes called the Ernst angle.

The third method for improving the sensitivity involves nuclear spin gymnastics. In all of these experiments, rather sophisticated pulse sequences are used in order to improve sensitivity. One of the first of these experiments was the driven equilibrium Fourier transform (DEFT) experiment (54, 55). In this early, rather primitive, experiment M_{xy} is first refocused in a spin-echo experiment in order to overcome the limitations of magnetic field homogeneity and then driven back to equilibrium in a time much shorter than that required by normal relaxation processes. Figure 6.1a shows the pulse timing scheme for this experiment; the time evolution of the magnetization is shown in Figure 6.1b

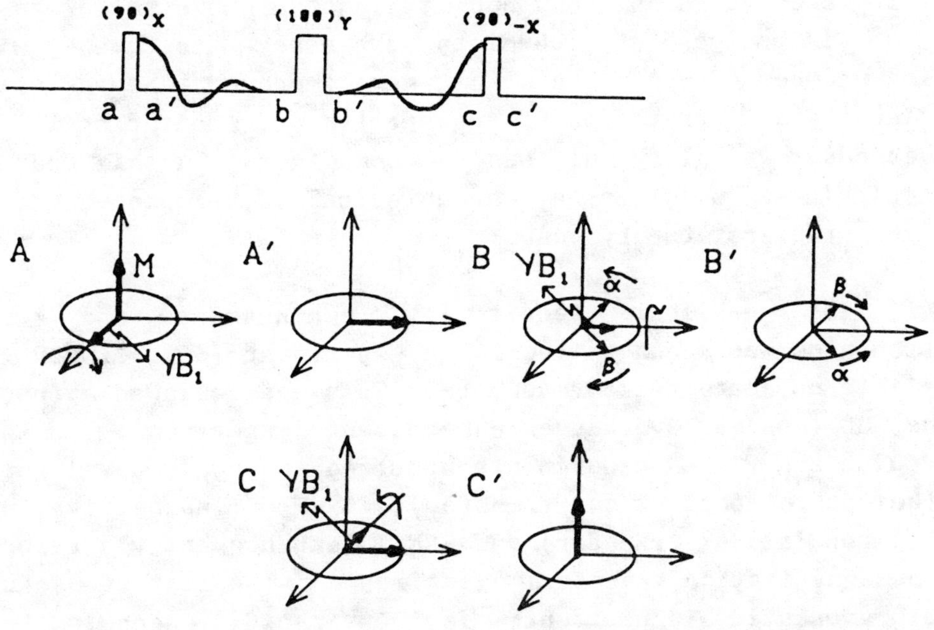

Figure 6.1 In the DEFT experiment, the signal is enhanced by refocusing the dephased magnetization. A $\pi/2$ pulse at (a) rotates the magnetization into the xy plane (a'). In the time interval τ between a' and b, the magnetization dephases due to field inhomogeneities. At (b) a π refocusing pulse is applied along the +y axis. At a time τ after the refocusing pulse, the magnetization is refocused along the +y axis (c). A $\pi/2$ pulse along the -x axis rotates the nearly undiminished magnetization back to its equilibrium position along the +z axis.

The DEFT method can give a sensitivity improvement of a factor of 40 or more for the condition that $T_1 \approx T_2 \gg T_2^*$. Since this condition was not usually met using the early rather primitive decoupling methods (e.g., using simple noise-modulated decoupling in a $^{13}C-\{^1H\}$ experiment $T_2 \ll T_1$), the DEFT and other early sensitivity enhancement methods were not often used. They also suffered from the facts that very accurate timing was required for the pulses and very homogeneous γB_1 fields were required. With the advent of more sophisticated r.f. electronics and pulse programmers, a number of more generally useful and powerful multipulse experiments were developed which have revolutionized multinuclear experiments with carbon-13, silicon-29, nitrogen-15, and so on.

6.2 The INDOR Experiment

NMR spectroscopy is probably the most powerful tool which a modern chemist has to study molecular structure and molecular dynamics, but it has been plagued from its inception with problems of low sensitivity. The early NMR spectroscopists frequently told their colleagues not to bother them for spectra unless they had "at least a bucketful of sample." But being an inventive lot, chemists have managed to come up with an almost endless number of experiments that have improved the sensitivity and simplified the interpretation of the spectra or, better still, have developed methods to simplify the spectra themselves. In this chapter we present a logical as well as a chronological development of these techniques over the last twenty years.

One of the earliest sensitivity enhancement and spectral interpretation aids was known as the INDOR experiment (56). INDOR is the acronym for internuclear double resonance. This was followed a few years later by the NOE (nuclear Overhauser enhancement experiment). With the advent of pulse and FT NMR experiments occurred the rapid development of the SPT (selective population transfer), INEPT (insensitive nuclei enhanced by population transfer) and ERNST (enhancement via refocussed nuclear spin-magnetization transfer) experiments. The list of useful and powerful spin gymnastics experiments is now rapidly becoming almost endless. We shall discuss the ones given above because they are relatively common, easy to execute, and provide substantial improvements in sensitivity and spectral simplification.

To describe clearly what happens in each of the above experiments it is most useful to consider the energy-level diagram for a simple spin system. We shall consider in detail the situation for a two-spin system and indicate how to extend it to the relatively simple three- and four-spin systems. We designate the spins A and X. For example, A could be a proton and X a carbon nucleus. In Figure 6.2a we show the energy levels for the case where the spin coupling J_{AX} between the two spins is zero, and in Figure 6.2b and c we show the energy levels for the case where J_{AX} is not zero. It is straightforward to show that the energy levels for this system in the absence of spin coupling are given by

$$E_4 = 1/2(v_a + v_x), \quad E_3 = 1/2(v_a - v_x),$$

$$E_2 = -1/2(v_a - v_x), \quad E_1 = -1/2(v_a + v_x), \qquad (6.1)$$

where v_a is the frequency of the a spin resonance and v_x is the frequency of the x spin resonance (e.g., 200 MHz for the protons and 50 MHz for the carbons). In the presence of spin coupling the energy levels are given by

$$E_4 = 1/2(v_a + v_x) + 1/4\ J_{ax}$$

$$E_3 = 1/2(v_a - v_x) - 1/4\ J_{ax}$$

$$E_2 = -1/2(v_a - v_x) - 1/4\ J_{ax}$$

$$E_1 = -1/2(v_a + v_x) + 1/4\ J_{ax} \qquad (6.2)$$

The spin label α in Figure 6.2 designates a spin aligned with the magnetic field (i.e., in the lower energy level), and the spin label β designates a spin aligned against the magnetic field. From quantum mechanics that the only transitions normally allowed in an NMR experiment are those for which the A spins change their spin quantum number by ± 1 or the X spins change by ± 1. Thus the transitions for the A spins are between levels 1 and 3 or between levels 2 and 4. Note that the X spins do not change their spin state between energy levels 1 and 3 or 2 and 4. Similarly, the transitions for the X spins are between levels 1 and 2 and between 3 and 4; the A spins do not change their spin state between these pairs of energy levels.

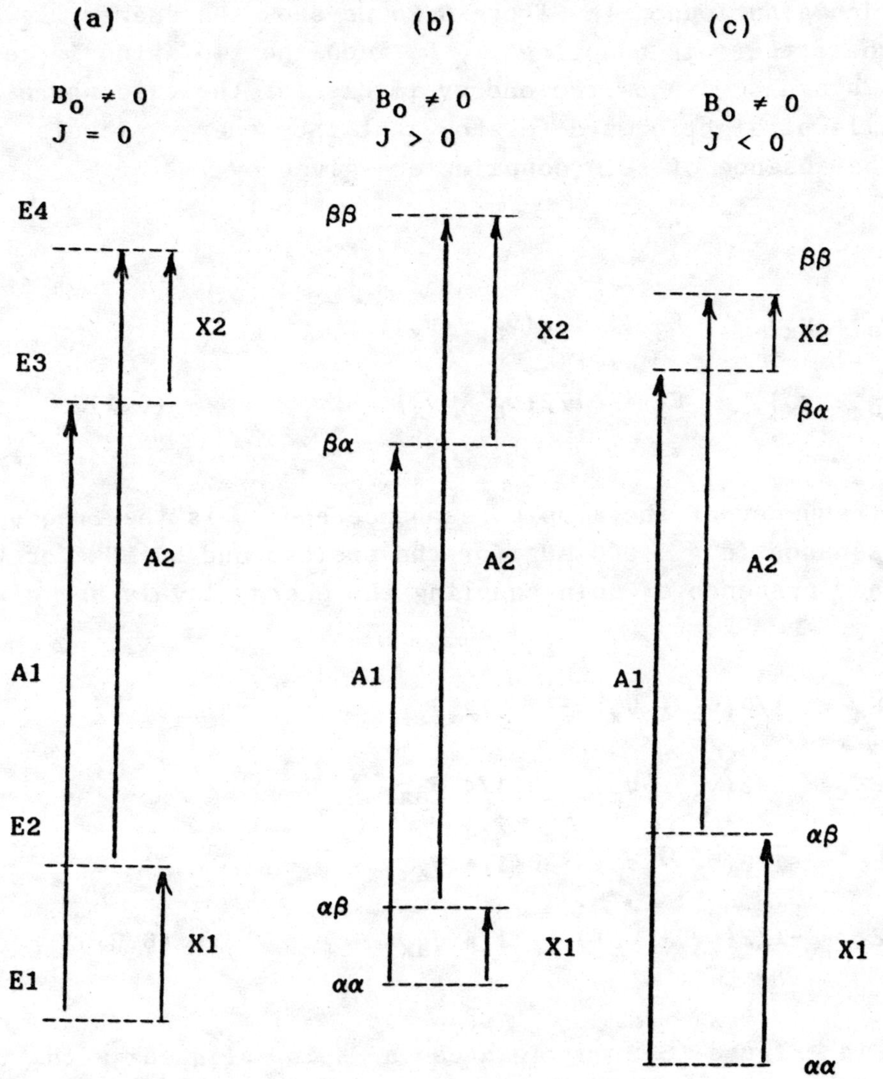

Figure 6.2 The energy level diagram for a two-spin system, AX. Both spins have a spin of 1/2. (a) In the presence of a dc magnetic field, B_o, but with a spin coupling constant, J_{AX}, of zero. (b) The energy levels for the case that $J_{AX} > 0$. Here the low-frequency X1 transition is between levels 1 and 2 and the low-frequency A1 transition is between levels 1 and 3. (c) The energy levels for the case $J_{AX} < 0$.

One is almost always be concerned only with the difference in populations between the various energy levels, since the intensity of the signals are directly proportional to these population differences. Let us then represent the small excess population of A spins in the energy levels by $+1/2n_a$, and the excess of X spins by $+1/2n_x$. In the same way,

In the same way we designate the small population depletions by $-1/2n_a$ and $-1/2n_x$. The relative population excesses or deficiencies of the four energy levels are then given by

$$P1 = +1/2\ n_a + 1/2\ n_x, \qquad P3 = -1/2\ n_a + 1/2\ n_x,$$

$$P2 = +1/2\ n_a - 1/2\ n_x, \qquad P4 = -1/2\ n_a - 1/2\ n_x. \tag{6.3}$$

The relative intensitities L(A1), L(A2), L(X1), and L(X2) of the spectral lines are given by the differences in the populations between these energy levels:

$$\begin{array}{llll} L(A1) = & 1/2\ n_a + 1/2\ n_x = P1 & L(A2) = & +1/2\ n_a - 1/2\ n_x = P2 \\ & -(-1/2\ n_a + 1/2\ n_x) = P3 & & -(-1/2\ n_a - 1/2\ n_x) = P4 \\ & \text{-------------------------}, & & \text{-------------------------}, \\ L(A1) = & n_a & LA(2) = & n_a \\ & & & \\ L(X1) = & 1/2\ n_a + 1/2\ n_x = P1 & L(X2) = & -1/2\ n_a + 1/2\ n_x = P \\ & -(\ 1/2\ n_a - 1/2\ n_x) = P2 & & -(-1/2\ n_a - 1/2\ n_x) = P4 \\ & \text{-------------------------}, & & \text{-------------------------}. \\ L(X1) = & n_x & L(X2) = & n_x \end{array} \tag{6.4}$$

Thus there are two spectral lines in the A (or proton) spectrum of equal intensity, and there are two lines in the X (or carbon) spectrum of equal intensity. However, since the gyromagnetic ratio of the protons is four times larger than the gyromagnetic ratio for the carbon nuclei, the proton population differences will be four times greater than for the carbons. We may express this more precisely by

$$n_a = (\gamma_a/\gamma_x)\ n_x, \tag{6.5}$$

which for protons and carbon gives

$$n_H \cong 4\ n_C. \tag{6.6}$$

The consequence of this equality is that the pair of lines in the proton spectrum are four times more intense than the pair of lines in the carbon spectrum. A schematic picture of the two pairs of spectral lines is given in Figure 6.3.

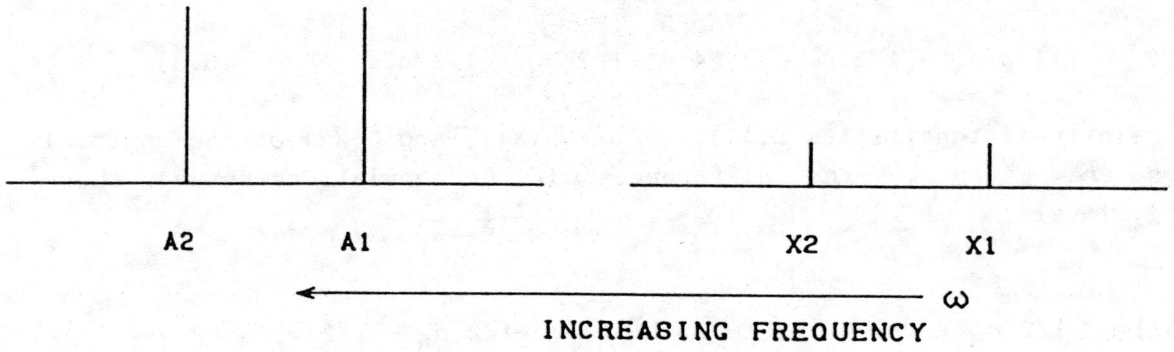

Figure 6.3 The two doublet spectra for an AX spin system. The spacing between each doublet is J_{AX}. For the AX system $(\delta_A - \delta_X) \gg J_{AX}$.

Let us now consider the INDOR experiment, one of the first attempts to obtain more information about energy levels and to obtain sensitivity enhancement. About fifteen years ago, this INDOR scheme was devised to obtain information about carbon-13 chemical shift values without having to directly observe the carbon resonance itself. Direct observation of the carbon resonance was very difficult using cw methods, because carbon-13 relaxation times are much longer than proton relaxation times, the natural abundance of carbon-13 is only 1 percent and carbon has a gyromagnetic ratio four times smaller than that for a proton. One can, however, readily observe the carbon-13 satellite lines with even an old, tired and not very sensitive proton NMR spectrometer, provided that the spectrometer is stable. Let us sweep the frequency of our proton NMR spectrometer such that we are precisely at the peak of one of the ^{13}C satellite lines in the proton NMR spectrum (see Figure 6.3). Now let us monitor the intensity of this line as we sweep through the ^{13}C spectrum with a high amount of r.f. power. As we pass through the resonance region of each of the lines in the carbon spectrum, we strongly perturb the carbon populations. That is, as the X1 line comes into resonance, the n_X values for the energy levels associated with that line will be decreased. Thus as we pass through the X1 line the populations of the energy levels are approximately given by:

$$P1 = +1/2\ n_a + 0\ ,$$

$$P2 = +1/2\ n_a - 0\ ,$$

$$P3 = -1/2\ n_a + 1/2\ n_x\ ,$$

$$P4 = -1/2\ n_a - 1/2\ n_x\ .$$

The proton line intensities are now approximately given by

$$\begin{aligned} L(A1) &= P1 - P3 \\ &= n_a - n_x \\ &= n_a - .25\ n_a \\ \hline L(A1) &= 0.75\ n_a \end{aligned} \quad , \qquad \begin{aligned} L(A2) &= P2 - P4 \\ &= n_a + n_x \\ &= n_a + 0.25 \\ \hline L(A2) &= 1.25\ n_a \end{aligned} \quad .$$

The result is that as one sweeps the B_2 r.f. field through, and perturbs the X1 line in the carbon spectrum, one causes a <u>decrease</u> in intensity of the A1 proton line and an <u>increase</u> in the intensity of the A2 proton line. It is simple to show that if we perturb the X2 carbon line, the opposite happens. We decrease the intensity of the proton A2 line and increase the intensity of the proton A1 line.

Let us recapitulate the experiment depicted in Figure 6.4. We monitor the intensity of the low-field ^{13}C satellite of the proton spectrum. We sweep the B_2 r.f. field through the carbon-13 resonance frequency with a large amount of r.f. power. As we pass through the X1 carbon line we observe a decrease in the intensity of the proton line (and we record the carbon-13 frequency at which this decrease occurs). We continue our sweep through the carbon-13 frequency region, and as we pass through the X2 line we observe an increase in the intensity of the proton line (and we record the frequency of the carbon-13 frequency at which this increase occurs). In this manner we are able to observe indirectly with very high precision the frequency of the two lines in the carbon-13 spectrum.

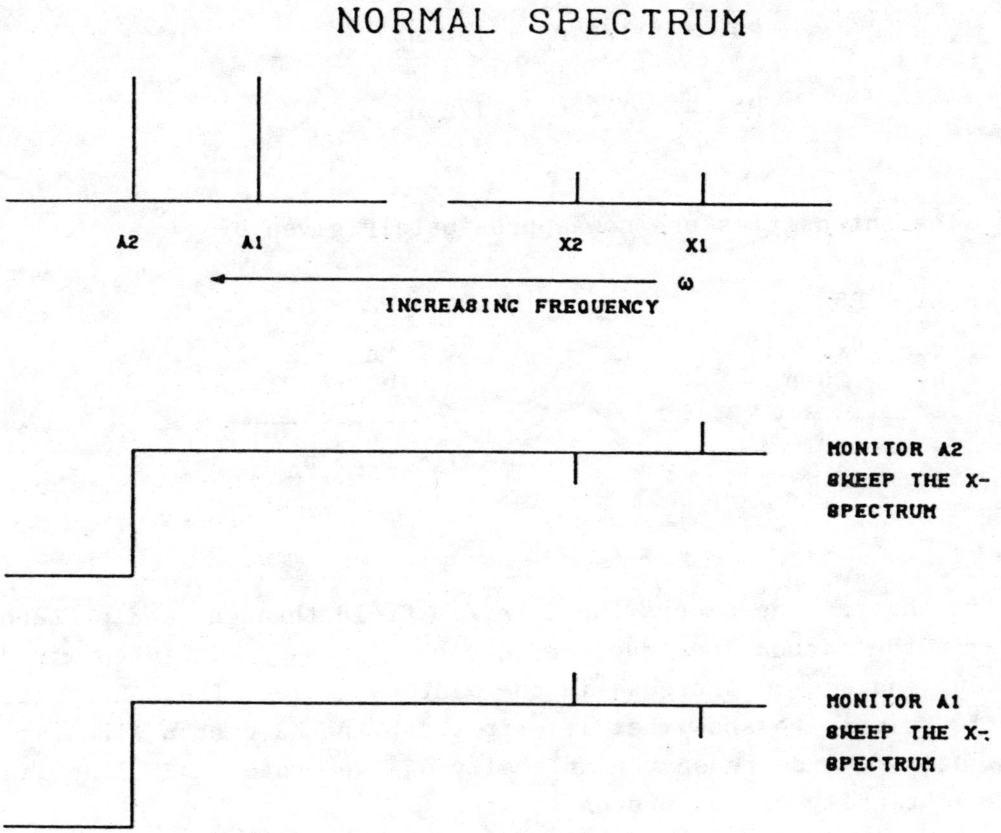

Figure 6.4 In an INDOR experiment the spectrometer observe frequency is set to coincide with the resonance frequency of the A2 (high-frequency) line in the A spectrum. That is, as we observe in the middle trace, one "parks" at the peak of the A2 line and then scans through the X spectrum at a relatively high power. As the two peaks in the X spectrum are traversed, the intensity of the A spectrum first decreases (as we pass through the resonance frequency for the X2 line) and then increases (as we pass through the resonance frequency for the X1 line). The reverse is true if we monitor the intensity of the A1 transition (lower trace).

6.3 The Nuclear Overhauser Experiment

Although the INDOR experiment is capable of providing useful information, it is a difficult experiment to execute and is very time consuming. With the steady improvement in spectrometers and the advent of high-resolution pulse NMR methods, carbon-13 NMR has become much more common, especially when broad-band proton decoupling is used (references 56a, b, c).

Broad-band proton decoupling is normally employed in ^{13}C NMR experiments for several reasons. Consider methyl iodide, CH_3I, as an example. Proton decoupling collapses the ^{13}C quartet (four lines with relative intensities of 1:3:3:1) to a single line with an intensity 24 times greater than the outer lines of the quartet. One might expect the 1:3:3:1 quartet to collapse to a single line with an intensity eight times greater than the outer lines. The additional threefold increase in sensitivity comes from the nuclear Overhauser enhancement (57). This is a cross-polarization phenomenon, and it is useful and still commonly used. Therefore, it is worthwhile to describe it in some detail.

Recall that for a single spin, I, relaxation from a non-equilibrium configuration is given by

$$dM_z(I)/dt = -R_I[M_z(I) - M_o(I)], \qquad (6.7)$$

where $M_z(I)$ = the z component of the magnetization of spin I at any given time, t.

$M_o(I)$ = the maximum, equilibrium magnetization of spin I,

R_I = $1/T_1$ for spin I.

If two spins, I and S, are spin coupled (e.g., I = proton and S = carbon in $CHCl_3$), then the relaxation behavior of spins I and S is interrelated and their relaxation can be described by two, coupled linear differential equations:

$$dM_z(I)/dt = -R_{II}[M_z(I) - M_o(I)] - R_{IS}[M_z(S) - M_o(S)], \qquad (6.8)$$

$$dM_z(S)/dt = -R_{IS}[M_z(I) - M_o(I)] - R_{SS}[M_z(S) - M_o(S)], \qquad (6.9)$$

where M_o is the equilibrium value for the magnetization.

In fact, the situation is a bit more complex since each of the lines in the ^{13}C or the ^1H doublet may relax at different rates. Happily, for most of the situations which we will encounter, it is reasonably valid to assume that all of the lines in a ^{13}C multiplet relax at the same rate. Now, if we saturate spin I (the proton spin in our example of $CHCl_3$), then $M_z(I) = 0$. If we wait for equilibrium conditions for spin S under these proton saturation conditions, then

$$dM(S)/dt = 0 = -R_{SI}[M_z(I) - M_o(I)] - R_{SS}[M_z(S) - M_o(S)] \qquad (6.10)$$

however, $M_z(I) = 0$, and at equilibrium $dM/dt=0$, so upon rearranging, 6.10 becomes

$$M_z(S) = M_o(S) + (R_{IS}/R_{SS})M_o(I). \qquad (6.11)$$

The result is that we have increased S spin magnetization, $M_z(S)$, by the amount $M_o(I)(R_{IS}/R_{SS})$. The nuclear Overhauser factor (NOF) and the nuclear Overhauser enhancement are defined by

$$NOF(S) = [M_z(S) - M_o(S)]/M_o(S), \qquad (6.12)$$

$$NOE(S) = NOF(S) + 1. \qquad (6.13)$$

If we substitute equation 6.11 into equation 6.12, we obtain

$$NOF(S) = \frac{M_o(S) + M_o(I)(R_{IS}/R_{SS}) - M_o(S)}{M_o(S)}$$

$$= \left[\frac{M_o(I)\ R_{IS}}{M_o(S)\ R_{SS}}\right]. \qquad (6.14)$$

Recall, however, that

$$M_o(S) = k\gamma_S S(S+1) \qquad (6.15a)$$

and

$$M_o(I) = k\gamma_I I(I+1). \qquad (6.15b)$$

Usually, for liquid samples when in the "extreme narrowing limit" (i.e., when $\omega_o\tau_c << 1$) we may write

$$R_{IS}/R_{SS} = S(S + 1)/[2I(I + 1)] \qquad (6.16)$$

and

$$NOF(S) = \left[\frac{M_o(I).R_{IS}}{M_o(S)\ R_{SS}}\right]$$

$$= \left[\frac{k\gamma_I I(I + 1)S(S + 1)}{k\gamma_S S(S + 1)\ 2I(I + 1)}\right],$$

$$NOF(S)_{max} = \gamma_I/2\gamma_S\ . \qquad (6.17)$$

$$NOE(S)_{max} = 1 + \gamma_I/2\gamma_S\ . \qquad (6.18)$$

For $I = {}^1H$ and $S = {}^{13}C$, recall that $\gamma_C = 4\gamma_H$ and consequently

$$NOF({}^{13}C)_{max} = 2.0\ ,$$

$$NOE({}^{13}C)_{max} = 3.0\ .$$

Thus, in proton-decoupled carbon-13 experiments we expect a three-fold enhancement of the NMR signals. However, we realize this maximum enhancement only if the following conditions hold:

1) The internuclear distance is short (about 1 Angstrom).
2) The relaxation is due solely to the nuclear dipole-dipole interaction.
3) The sample is in the extreme narrowing limit ($\omega_o\tau_c<<1$).

For carbon nuclei with directly bonded hydrogens, condition 1 above is always met. For a surprisingly large number of cases, however, the carbon (or nitrogen, or selenium, or "other" spin-S nuclei) relaxation is not due entirely to dipole-dipole relaxation and condition 2 above is not met. For example, the spin-rotation interaction usually makes a substantial contribution to the relaxation of the methyl carbons. Dissolved oxygen in the sample also contributes to the relaxation via the unpaired electrons on the oxygen molecule. For nuclei with a negative gyromagnetic ratio, such as ^{15}N or ^{29}Si, the problems can be especially troublesome since the NOE gives a negative enhancement. In this situation the NOE can, and often does, result in a signal degradation.

Although the most common use of the NOE experiment is for sensitivity enhancement in $^{13}C<^{1}H>$, proton-decoupled ^{13}C experiments, it is often used for homonuclear $^{1}H<^{1}H>$ decoupling experiments. In the homonuclear $^{1}H<^{1}H>$ experiment the maximum enhancement predicted from equation 6.18 is a factor of 1.5, that is, a 50 percent enhancement. It is beyond the scope of this brief treatise to derive the details, but it can be shown that where $\omega_o\tau_c << 1$ and for nuclear dipolar relaxation, two interacting protons A and B (or groups of equivalent protons) can be described thus:

$$R(AB)/[R(BB)] = 2 + kr(AX)^6 , \qquad (6.19)$$

where

R(BB) = total inter- and intramolecular relaxation rate contributions to proton B,

R(AB) = specific contribution to the relaxation of proton B arising from proton A,

r(AX) = the internuclear distance between protons A and B,

k = an empirically determined constant,

and

$$NOE(^1H) = 1/2, \quad \text{for } r(AB) < r^o(AB), \tag{6.20}$$

$$NOE(^1H) = kr(AB)^{-6}, \quad \text{for } r(AB) > r^o(AB), \tag{6.21}$$

where r^o is an empirically determined proton-proton internuclear distance (typically $r^o \approx 1.9$ Angstrom).

The details of this derivation and a number of examples are given in a paper by Bell and Saunders (58). Using Dreiding models and empirical results, these authors obtain plots for the H-H and H-CH_3 interactions shown in Figure 6.5.

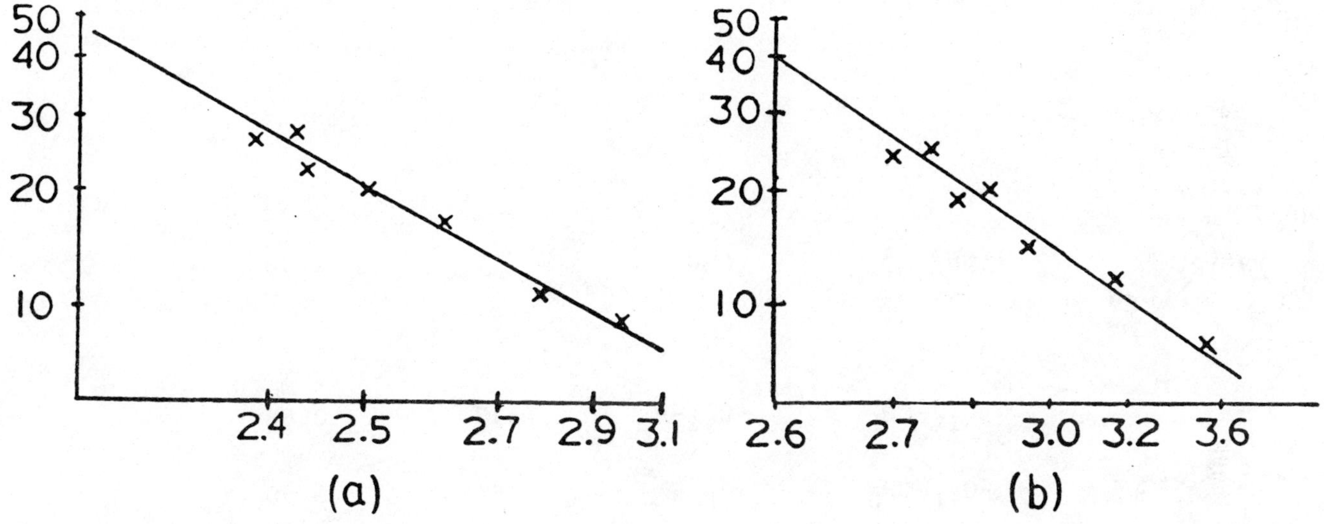

Figure 6.5 Plots of the proton-proton NOE as a function of the distance between the interacting proton groups. In (a) the percent enhancement is plotted as a function of the distance between different protons. In (b) the percent enhancement is plotted as a function of the distance between a proton and the methyl protons with which it is interacting. For more details see reference 58.

Some typical examples of some of the molecules studied are shown in Table 6.1.

Table 6.1

MOLECULE	IRRADIATED GROUP	OBSERVED GROUP	PERCENT NOE	BOND DISTANCE R_{AX} (A)
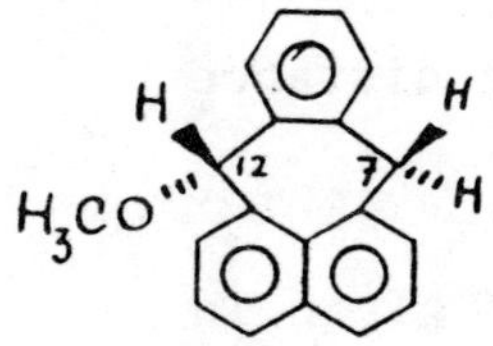	C_7-H (axial)	C_{12}-H	28	2.40
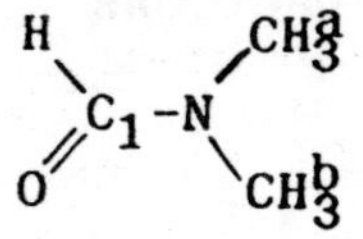	CH_3^a	C_1-H	28	2.69
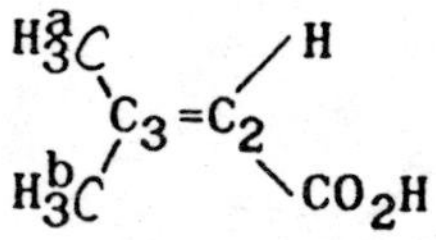	CH_3^a	C_2H	17	2.90
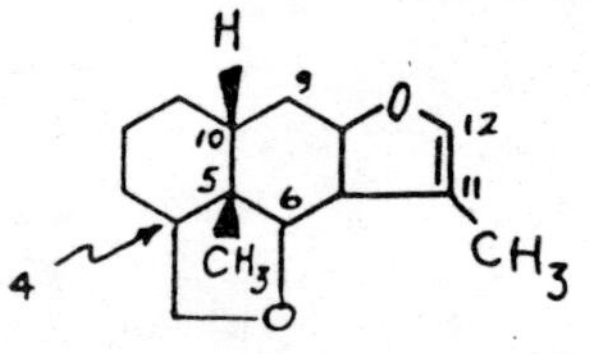	C_{11}-CH_3	C_{12}-H	17	2.88
	C_{11}-CH_3	C_6-H	6	3.56
	C_5-CH_3	C_6-H	23	2.74

When more than one proton or group of nonequivalent protons relax a single proton, it is not generally possible to determine each interuclear distance uniquely, but only their ratios. Although this can be a most useful technique, it is important that very great care be used in the preparation of the sample and in the execution of the experiment:

a) All paramagnetic materials must be removed (especially dissolved oxygen.
b) All solid particles should be removed by filtration.
c) At least 5 freeze-pump-thaw cycles should be used to insure that all oxygen is removed.
d) The B_1 (observe) and B_2(irradiate) r.f. fields must not saturate the observed resonance.
e) The frequency and power level of the irradiation frequency, B_2, must be carefully and precisely adjusted.

For more details the reader is referred to Chapter 3 of the excellent book by Noggle and Schirmer (59), which gives a comprehensive and clear treatment of the nuclear Overhauser experiment.

6.3 The Selective Population Transfer (SPT) Experiment or Selective Population Inversion (SPI) Experiment

NOE experiments have provided us with a wealth of information and a much needed and welcome improvement in sensitivity, especially for ^{13}C experiments. By making use of gated decoupling experiments it is possible to obtain ^{13}C spectra with or without proton decoupling and with or without the nuclear Overhauser enhancement. Useful though the NOE experiments are, they suffer from a number of drawbacks. For example, in the $^{13}C\langle^{1}H\rangle$ NOE experiment we have the following problems:

1) The ^{13}C signal intensities vary not only as a function of the relative number of chemically equivalent carbon nuclei present, but they also vary as the NOE values vary for those carbon nuclei; the NOE values can range from 1 to 3.

2) The ^{13}C intensities vary as a function of the T_1 values for the various carbon nuclei.

3) The NOE experiment is not selective.

4) The pulse repetition rate depends upon the ^{13}C relaxation times,which, in general, are rather long.

Before we discuss in detail the sensitivity enhancement experiments which follow, it is important to have a clear understanding of the energy-level diagrams, of how the spins are labeled, and of how populations among the energy levels can be changed. Consider the AX spin system (A, for example, could be a proton and X could be a carbon, as in $^{13}CHCl_3$), whose energy-level diagram for the normal equilibrium situation is shown in Figure 6.6a and whose spectrum is depicted in Figure 6.6c. Here, as before, the populations of the energy levels are given by

$$P1 = \rho + 1/2\, n_a + 1/2\, n_x \qquad P3 = \rho - 1/2\, n_a + 1/2\, n_x$$
$$P2 = \rho + 1/2\, n_a - 1/2\, n_x \qquad P4 = \rho - 1/2\, n_a - 1/2\, n_x$$

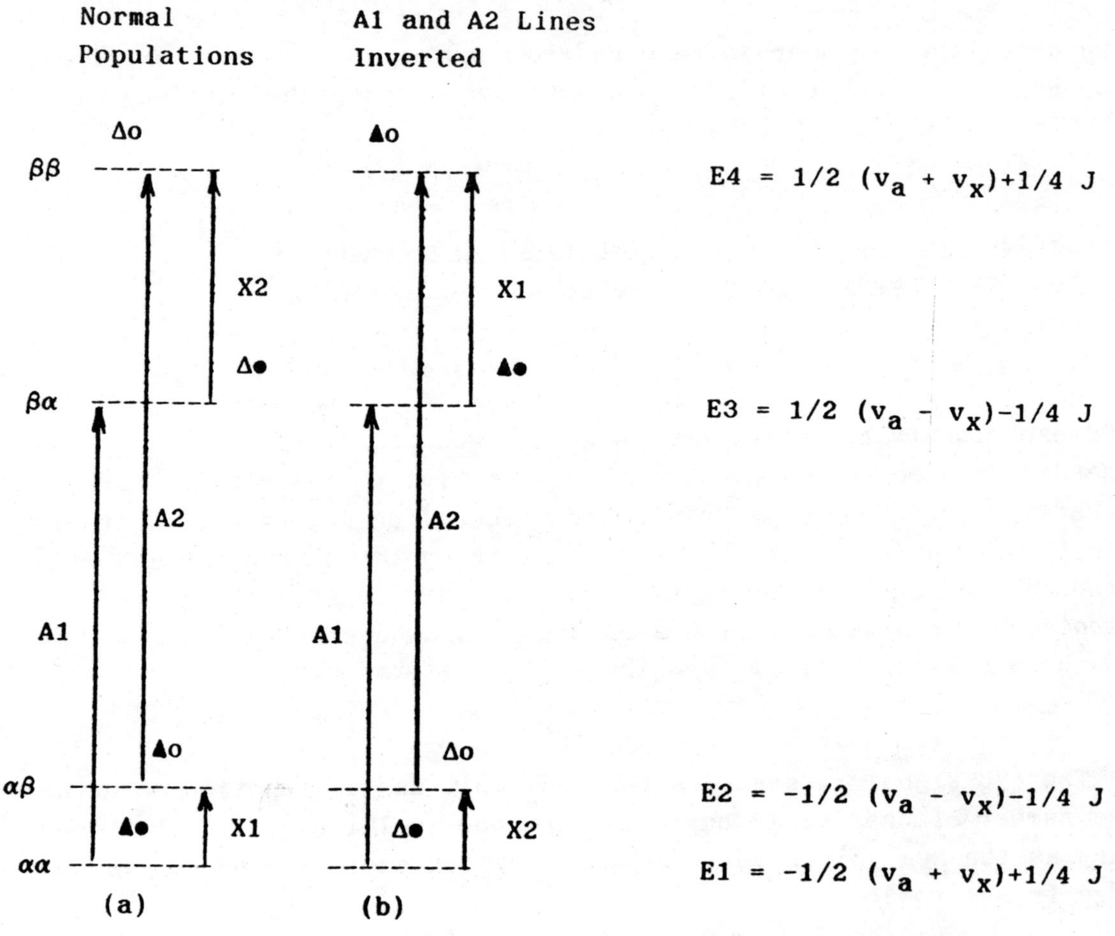

▲ = excess A spin population
Δ = deficit A spins population
● = excess X spin population
o = deficit X spin population

Figure 6.6a, b. The energy-level diagram for an AX spin system with a positive spin-coupling constant. The value of the energies and the relative populations of the various levels are indicated.

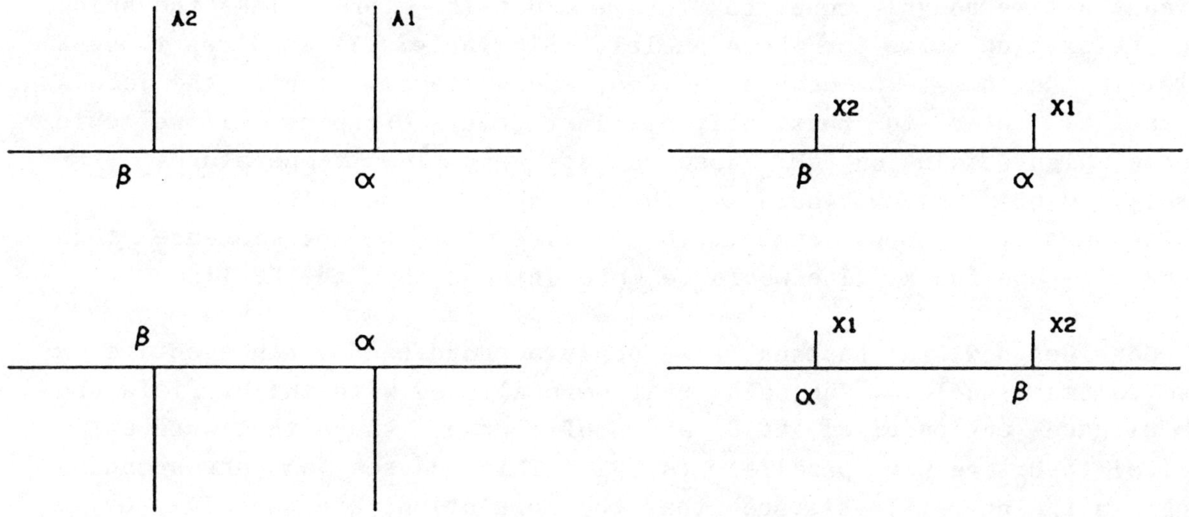

Figure 6.6c, d. (c) The normal A and X spectra for the two-spin AX system. (d) Both of the lines in the A spectrum are inverted. This changes the relative populations of the energy levels (see 6.6b), but does not change the appearance of the X spectrum. Note: The spectra and the energy levels are for the case J_{AX} is positive.

The transition that we label A1 in the A spectrum and in the energy-level diagram involves a transition from the most highly populated state (state 1) to a state of lower population (in this case state 3). Note that the A1 label is for a physically distinct group of spins, namely, those nucleus A spins which are in the lowest energy level and consequently have a spin label α. That is, they represent the approximately 1/2 fraction of the A nuclear spins which are spin coupled to those X nuclei which have spin α. These spins will remain in level 1 for at least a time roughly equal to $3T_2(A)$, that is, three times the spin-spin relaxation time for the A nuclei. Similarly, X1 involves those X nuclei in the lowest, or most populated, energy level. Thus, the labels A1 and X1 refer to physically distinct groups of spins. If we could observe high-resolution NMR spectra at very low temperatures (for example, 0.001^{o} K) we would see that because of the differences in the population of the energy levels, the A1 line would be more intense than the A2 line and the X1 line would be more intense than the X2 line.

Consider now what happens if we apply a broad-band, nonselective π pulse to the A nuclei. The spins that were aligned with the B_o field are now aligned antiparallel to B_o and, of course, those that were antiparallel to B_o are now parallel to B_o. This causes a corresponding change in the energy levels such that the populations are now given by

$$P1 = \rho - 1/2\ n_a + 1/2\ n_x, \qquad P3 = \rho + 1/2\ n_a + 1/2\ n_x\ .$$
$$P2 = \rho - 1/2\ n_a - 1/2\ n_x, \qquad P4 = \rho + 1/2\ n_a - 1/2\ n_x\ .$$

If we use a $\pi/2$ pulse at this point to observe the state of the spins, we obtain the spectra shown in Figure 6.6d. As expected, the A1 spins have simply been inverted. The low-frequency line, A1, is still associated with the most highly populated spin state; these spins are still coupled to X spins in the β spin state. At first glance it would appear that little has happened to the X part of the spectrum. A more careful inspection shows that the labels on the two lines have been reversed. Remember that X1 labels the transition associated with the most populated spin state, which is now the high-frequency transition. This line is still associated with A nuclei in the α spin state. The X2 transition, associated with the level of lower population, is now the low-frequency transition, as indicated in Figure 6.6d.

The result is that spin labels α and β, which designate the spin state of the A spins to which the X spins are coupled, are reversed. The high-frequency X transition is now coupled to an A spin in the α state and the low-frequency X transition is now coupled to an A spin in the β state (see Figure 6.6d). This subtle change is very important and has powerful consequences in the INEPT and ERNST experiments, which are discussed in the next sections.

Over the past several years a number of new experimental procedures using pulse NMR methods have been devised which are much superior to the NOE experiments. One of the earliest of these experiments was the selective population transfer, or SPT, experiment (60). This experiment is sometimes called the SPI experiment (selective population inversion), and it is very similar to the INDOR experiment. Consider again $CHCl_3$ as an example of the IS (or AX) two-spin system. The proton spectrum and the carbon spectrum are shown in Figure 6.6c. We shall refer to the two spins as the A (or the proton) spin and the X (or the carbon) spin.

In this experiment we invert one of the proton satellite spectral lines (the large, intense singlet due to nuclei containing carbon-12 is of no interest to us in this experiment) using a "soft" or selective proton pulse, and we then observe the carbon-13 spectrum. Recall, (equation 2.1), that $\theta = \gamma H_1 t_p$, so under the conditions

$$\gamma H_1 = 25 \text{ Hertz},$$
$$\omega_o = \omega_\alpha ,$$
$$t_p = 2.0 \text{ msecs},$$

the magnetization vector associated with the α line in the proton spectrum will be rotated by 180° and the magnetization associated with the β line in the spectrum will be left unperturbed. The α transition arises from those protons which are coupled to carbon-13 nuclei which have a spin quantum number of + 1/2. In Figure 6.7, this is the H1 transition. If we invert the magnetization associated with this transition, the energy-level diagram changes to that shown in Figure 6.7b.

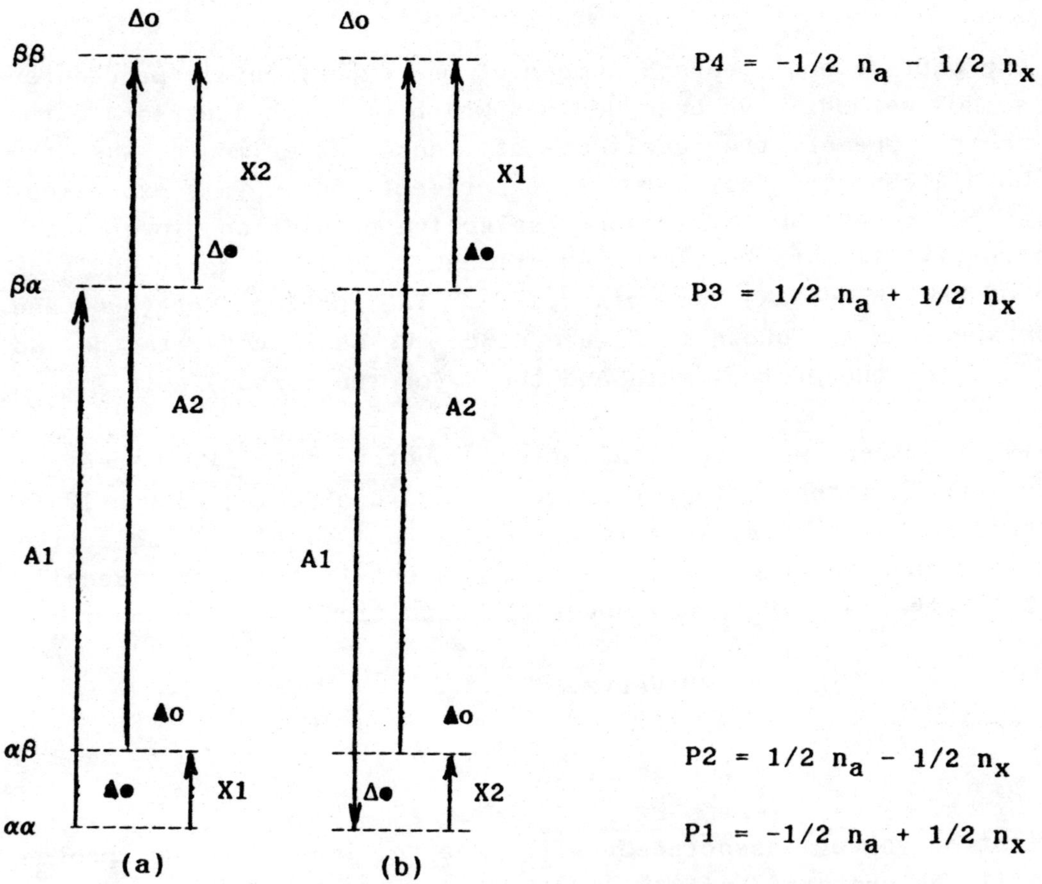

▲ = excess A spin population
Δ = deficit A spins population

● = excess X spin population
o = deficit X spin population

Figure 6.7 (a) The normal energy-level diagram for an AX spin system. (b) If only the A1 line in the A spectrum is inverted, only the populations in energy levels 1 and 3 will be interchanged.

In the same manner as before, we can calculate the new intensities for the lines in the carbon-13 spectrum:

$$L(C1) = P1-P2 = -1/2\ n_H + 1/2\ n_C - (+1/2\ n_H - 1/2\ n_C)$$

$$L(C1) = -n_H + n_C = -3n_C \qquad (6.22)$$

Since again $n_C = 4\ n_H$, $L(C1) = -3n_C$ and, similarly,

$$L(C2) = P3 - P4 = 1/2\ n_H + 1/2\ n_C\ (-1/2\ n_H - 1/2\ n_C)$$

$$L(C2) = n_H + n_C = 5\ n_C \qquad (6.23)$$

Thus, by inverting the α line in the proton spectrum and then observing the carbon spectrum, we see not a 1:1 doublet, but a -3:+5 doublet. The SPT experiment for CH_2Cl_2 (the A_2X case) similarly gives not a 1:2:1 triplet, but a -7:+2:+9 triplet. In general, the normal multiplet intensity ratios given by Pascal's triangle (Figure 6.8a) are, in the SPT experiment, given by the modified Pascal's triangle shown in Figure 6.8b.

(a)	(b)	
1	1	A
1 1	-3 5	AX
1 2 1	-7 2 9	A_2X
1 3 3 1	-11 -9 15 13	A_3X
1 4 6 4 1	-15 -28 6 36 17	A_4X
1 5 10 10 5 1	-19 -55 -30 65 50 21	A_5X
1 6 15 20 15 6 1	-23 -90 -105 20 135 102 25	A_6X

Figure 6.8 (a) Pascal's triangle shows the relative intensities for the multiplets in an NMR spectrum, e.g., a 1:1 doublet, a 1:2:1 triplet, a 1:3:3:1 quartet, and so on. For the SPI (or SPT) experiment, the relative intensities are changed to that shown in (b), e.g., a -11:-9:15:13 quartet rather than a 1:3:3:1 quartet.

The SPT, or SPI, experiment is similar to the INDOR experiment in that it involves changing the populations among the various energy levels. In the INDOR experiment some of the carbon magnetization goes to zero and one changes the proton signal intensities. In the SPT experiment we invert the proton population for some of the energy levels and observe the carbon-13 spectrum. Although this model is a bit naive, it does predict quite accurately the experimental observations, and allows one to make accurate predictions of the observed intensities. Spectra for samples of $CHCl_3$, CH_3I, and $(CH_3)_2CO$ are shown in Figure 6.9a-c. As can be seen, the sensitivity enhancements obtained are impressive. Unfortunately, there are a number of difficult aspects to this experiment:

a) One must know the precise positions of the carbon-13 satellite lines in the proton NMR experiment.

b) A "soft" pulse exactly centered on the chosen satellite line must be used.

c) Separate experiments must be executed for each of the satellite lines.

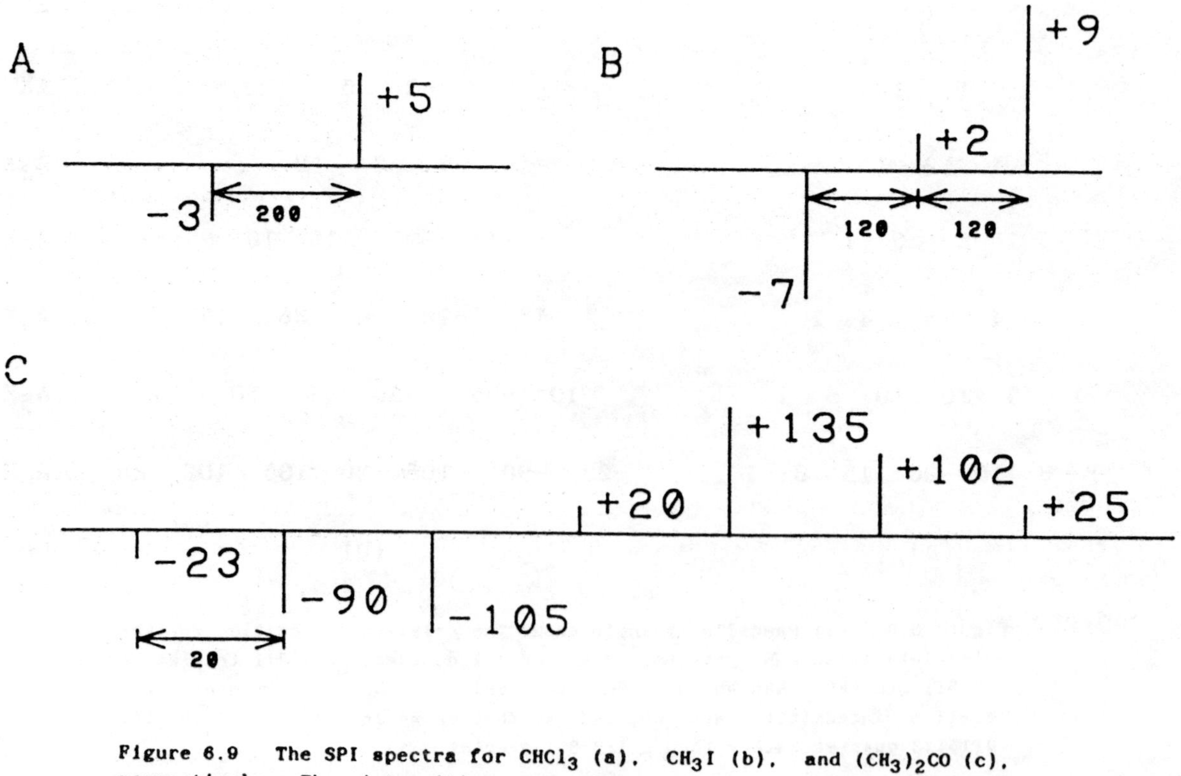

Figure 6.9 The SPI spectra for $CHCl_3$ (a), CH_3I (b), and $(CH_3)_2CO$ (c), respectively. The intensities are those predicted by the SPI Pascal triangle shown in Figure 6.7a.

6.4 The INEPT Experiment

Fortunately, a much more elegant and versatile experiment is available which can provide the same information as the SPT or SPI experiment, but with much greater ease. This experiment, the INEPT experiment (61-65), uses short, powerful nonselective pulses. It is an easy experiment to execute and is very forgiving in terms of the various experimental errors that can occur in setting pulse widths, r.f. phases, and so on.

In the INDOR experiment we destroy some of the magnetization of one spin (e.g. the ^{13}C spin) while monitoring the line intensity of a coupled, second spin (a proton). In the NOE experiment we transfer some of the magnetization of an abundant, high-γ spin to a less abundant, low-γ spin. In the SPT experiment we observe changes in the line intensities of one spin (e.g., a carbon spin) which result when we rearrange the distribution of the spin populations among the energy levels. Note that in this latter experiment we do not change the total net magnetization of the coupled spin system or the individual spins. Of these experiments the SPT experiment contains the most information. An experimentally more useful form of this experiment is the INEPT experiment (insensitive nuclei enhanced by polarization transfer).

Consider once again our old friend $CHCl_3$ as an example of a two-spin system (where A = proton and X = carbon-13). In this experiment we subject the carbon and proton spins to a carefully planned sequence of pulses which cause the spins, in turn, to go through a number of useful spin gymnastics. Figure 6.10 shows the pulse sequences for the carbon and proton spins.

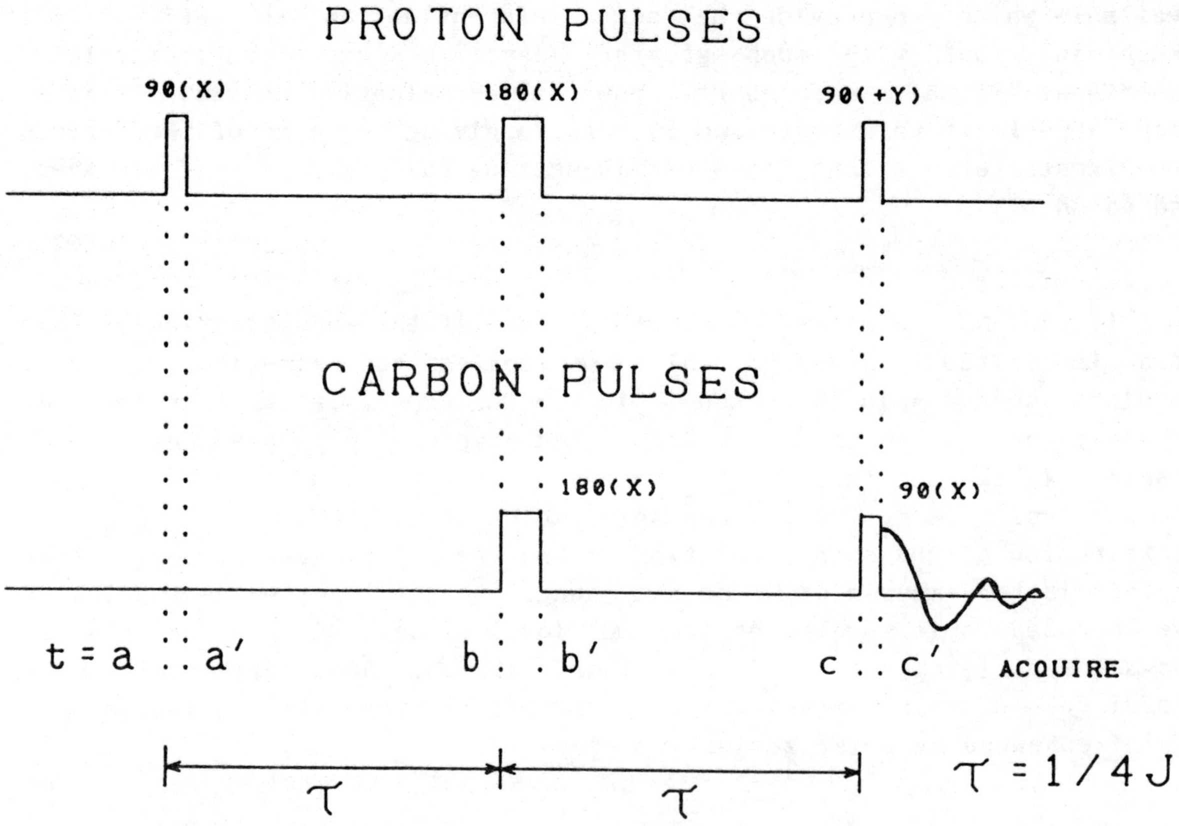

Figure 6.10 The pulse sequences for the A and X spins in an INEPT experiment. The time between points a' and b and between b' and c is τ sec, where $\tau = 1/(4J)$ sec. The final $\pi/2$ pulse for the A spins is phase shifted by 90° so that it lies along the + y axis.

Now let us follow in some detail the motions of the spins during this sequence of pulses. In order to keep our vector diagrams as simple as possible, we shall treat our system as though ω_0, the transmitter operating frequency, were placed precisely at the center of the multiplet. This causes no loss of generality, since any dephasing caused by chemical shift effects is refocused by the π-pulses. Consider first the motion of the protons. A $\pi/2(+x)$ pulse tips the A (proton) magnetization along the +y axis in the rotating frame (see a' in Figure 6.10). Since the transmitter reference frequency is positioned between the two lines in the proton doublet spectrum, the magnetizations for the two lines will precess at the same relative rate in the rotating frame, but in opposite directions. The higher-frequency proton line is the one which is spin coupled to a carbon in the α spin state. We designate this proton magnetization as the α magnetization (see Figure 6.11). Similarly, the lower-frequency line in the proton spectrum is the one which is spin coupled to the carbon β spin. We designate this proton magnetization as the β magnetization. Since the β magnetization is at a lower frequency than the r.f reference frequency, it will dephase in a direction opposite that of the α magnetization.

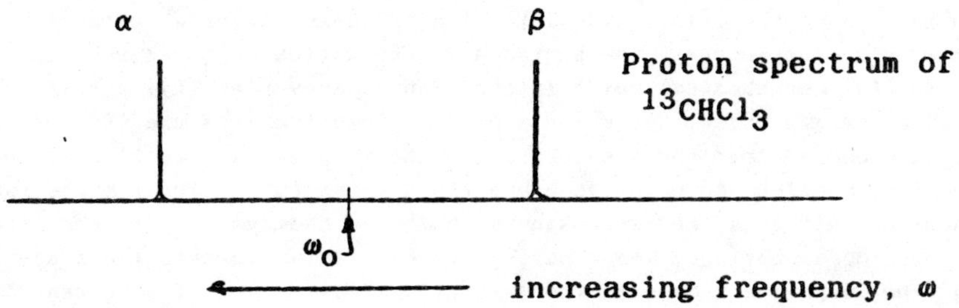

Figure 6.11 The NMR spectrum for the A spins in an AX system. The transmitter frequency, ω_0, is positioned between the doublet lines. This is done in order to simplify the vector diagrams. It is not essential, since the π pulses used in these experiments will refocus any dephasing caused by the chemical shift.

At time $t = b = 1/4J$, the proton α and β magnetizations will be 90° out of phase with one another (see b in Figure 6.12). If we apply a $\pi(+x)$ pulse to the protons they will be rotated 180° about the x axis. If at the same time we apply a $\pi(+x)$ pulse to the carbon spins, we invert the carbon magnetization and also interchange the spin labels on the proton magnetization. Thus, after this double π pulse the spins are as

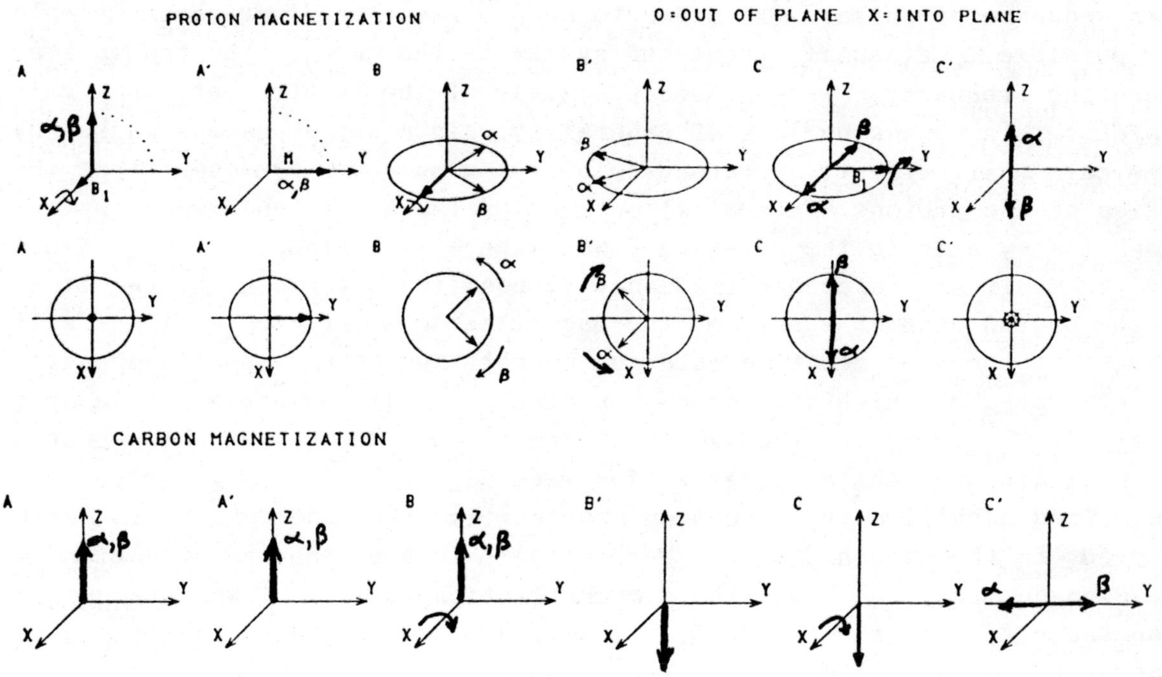

Figure 6.12 The evolution of the magnetizations for the A and X spins in an INEPT experiment. A is usually taken to be a proton spin and X is taken to be a carbon spin. At point (a) a $\pi/2$ pulse inverts both lines in the proton spectrum. The proton A magnetization dephases because of spin coupling and chemical shift interactions. At a time (b), $\tau = 1/(4J)$ sec after the $\pi/2$ pulse, the α and β proton magnetizations are 90° out of phase with one another and π pulses along the x axes in each rotating frame are applied to both the A and the X spins (b'). The π pulse for the A spins initiates the refocusing of both the chemical shift and the spin-coupling dephasing. The π pulse for the X spins inverts the X spins along the negative z axis in the X rotating frame, and it also cancels the effect of the A spin refocusing pulse for the spin-coupling interaction. Since the spin states of the X spins are interchanged by the π pulse for the X spins, the spin labels on the A spins are interchanged. This causes the A spin-coupling dephasing to continue while the A chemical shift dephasing is refocused. At point (c), which is τ seconds after the π pulses, the chemical shift dephasing of the A spins is refocused, but the α and β magnetizations of the A spins are now 180° out of phase because of the spin-coupling dephasing. Thus at (c) the α and β spins of the A nucleus are antiparallel along the +x and -x axes in the A spin rotating frame. The $\pi/2$ pulse on the A spins rotates them about the +y axis such that they are now antiparallel along the +z and -z axes. This also orders the X spins into an antiparallel state. The $\pi/2$ pulse along the x axis for the X spins rotates them into the xy plane and signal acquisition for the X spins is initiated. Note that the configuration of the A and X spins is identical to that in an SPI experiment.

shown in b' in Figure 6.12. The π pulse on the protons will cause the proton spin dephasing of the chemical shift and spin-coupling effects to refocus. The π pulse on the carbon nuclei will not affect the refocusing of the proton chemical shifts, but since it reverses the α and β proton spin labels, it nulls the refocusing of the spin-coupling effects of the proton dephasing. That is, the dephasing of the proton spins due to spin coupling will continue. We now wait a further time, $t = 1/4J$. Since the proton α and β magnetizations did not refocus but continued to dephase, at time $t = 2/4J = c$ the proton α and β magnetizations are antiparallel along the +x and -x axes, respectively. A further $\pi/2(+y)$ pulse rotates the proton magnetization into an antiparallel configuration along the +z and -z axes as shown in c' in Figure 6.12, and a $\pi/2(+x)$ ^{13}C pulse rotates the carbon magnetization into the xy plane. At this point, c', we begin acquisition of the carbon-13 NMR signal.

This sequence of pulses gives a result which for $CHCl_3$ is identical to the SPI experiment. It has the advantage that it is much easier to execute and results in an enhanced carbon-13 signal, just as in the SPI experiment. For samples with more than one carbon resonance, the signal is enhanced for all of the carbon spins. In contrast, the SPI experiment must be repeated for each different carbon resonance, since highly selective proton pulses are used to invert the magnetization.

6.5 The ERNST Experiment

The INEPT experiment is a powerful and useful experiment and affords the spectroscopist a great deal of information even under conditions where NOE experiments would be very difficult or impossible. The INEPT experiment does suffer from one major disadvantage: it often provides too much information. In the INEPT experiment, proton coupling is always present. For a molecule with a large number of carbons the spectra can become complex due to all the proton-coupled fine structure. If the proton-decoupling transmitter is turned on at time t = c' (see c' in Figure 6.10), in order to simplify the spectrum by suppressing the fine structure due to spin coupling, the carbon signal simply disappears. Without the proton spin coupling the two antiphase components of the carbon magnetization (see c' in Figure 6.10) remain 180° out of phase in the xy plane and the two NMR signals generated by their precession exactly cancel. The ERNST experiment (66) allows the carbon spins to refocus, and then proton decoupling produces a standard carbon-13 NMR spectrum with the expected enhancement. The pulse timing sequence for the ERNST experiment is shown in Figure 6.13.

Up to time t = c' (see c' in Figures 6.10 and 6.13) the ERNST and INEPT experiments are exactly the same. Two further evolution times of $\Delta = 1/4J$ allow the carbon magnetizations to refocus, this time along the x axis. Here, as in previous experiments, the function of the π pulses on the protons is to refocus the dephasing of both chemical shift and spin coupling. The π pulses on the carbon spins counteract the refocusing for the proton spin coupling (by reversing the α and β spin labels), and during the Δ evolution times, the π carbon pulses refocus any carbon chemical shift dephasing. The rotating frame pictures for the ERNST experiment are given in Figure 6.14.

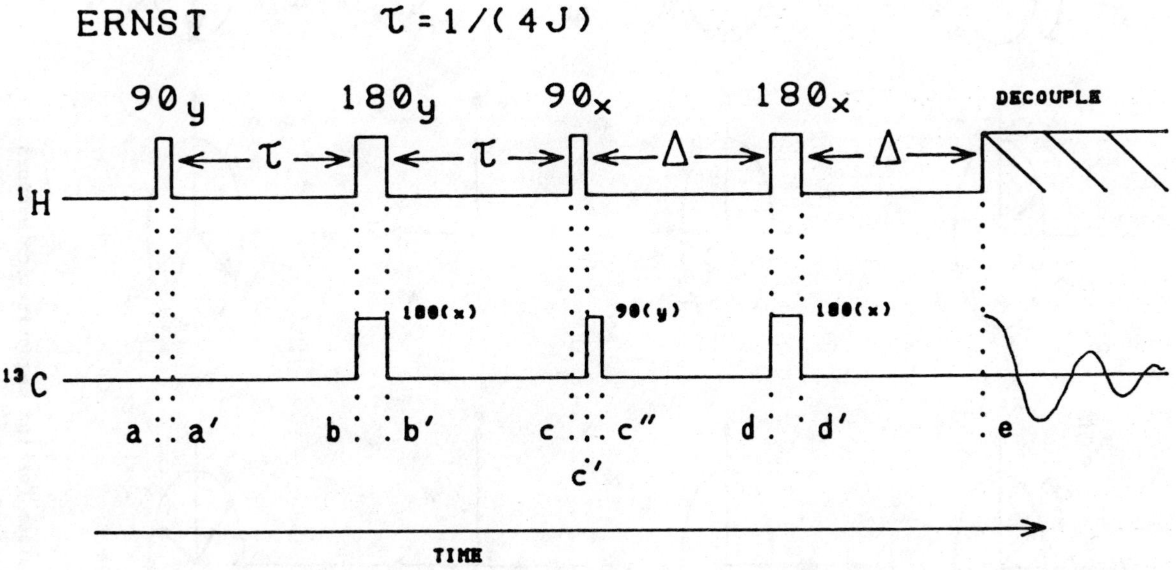

Figure 6.13 Timing sequence for the carbon observe-transmitter and proton decoupling channels for an ERNST cross-polarization experiment. The first three pulses for the protons and the the first two pulses for the carbon nuclei are the same as in the INEPT experiment. The Δ evolution times and the additional π pulses cause the carbon magnetizations to refocus. A carbon magnetization can then be observed in the presence of proton decoupling.

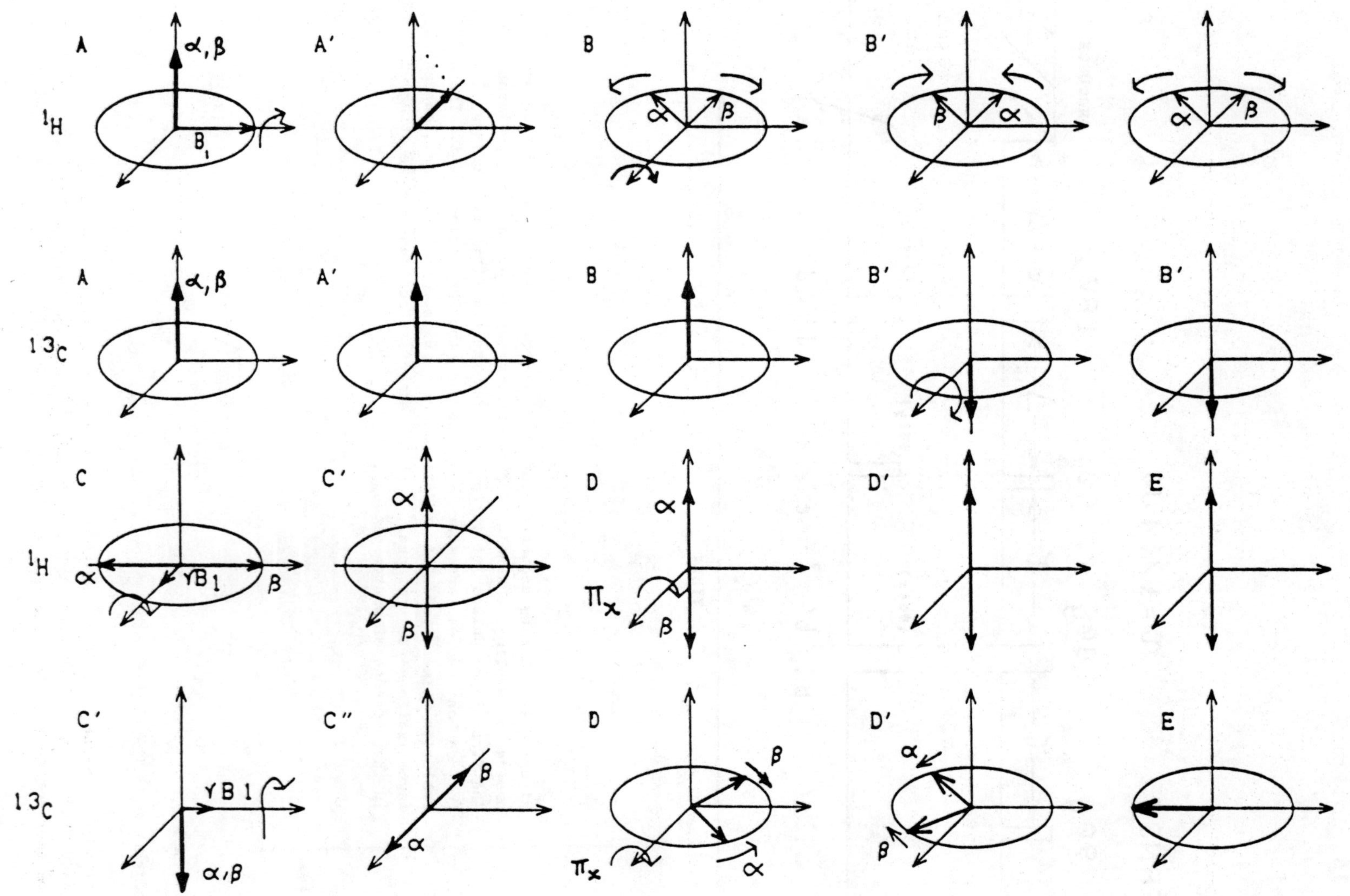

Figure 6.14 Rotating frame representation for the magnetization vectors of nuclei A and X during an ERNST experiment.

Up to this point we have considered polarization transfer experiments only for the case of two spins, A and X, both with a spin of 1/2. It is straightforward to show (refrence 66) that the enhancement is a function of both the spin coupling constant J_{AX} and the spin multiplicity. In particular, for the AX, A_2X and A_3X cases the enhancements E(AX), $E(A_2X)$ and $E(A_3X)$, respectively, are given by:

$$E(AX) = (\gamma_A/\gamma_X)\ \sin(2\pi J\tau)\ \sin(2\pi J\Delta), \qquad (6.24a)$$

$$E(A_2X) = (\gamma_A/\gamma_X)\ \sin(2\pi J\tau)\ \sin(4\pi J\Delta), \qquad (6.24b)$$

$$E(A_3X) = (3\gamma_A/4\gamma_X)\ \sin(2\pi J\tau)[\sin(2\pi J\Delta) + \sin(6\pi J\Delta)]. \qquad (6.24c)$$

In practice, τ is adjusted such that $\sin(2\pi J\tau) = \sin(\pi/2) = 1$, which gives the maximum enhancement. This is not a restrictive condition, since the $\sin(2\pi J\tau)$ term is related to the proton dephasing, which is independent of the number of protons attached to a given carbon nucleus. Plots of the enhancements are given in Figure 6.15.

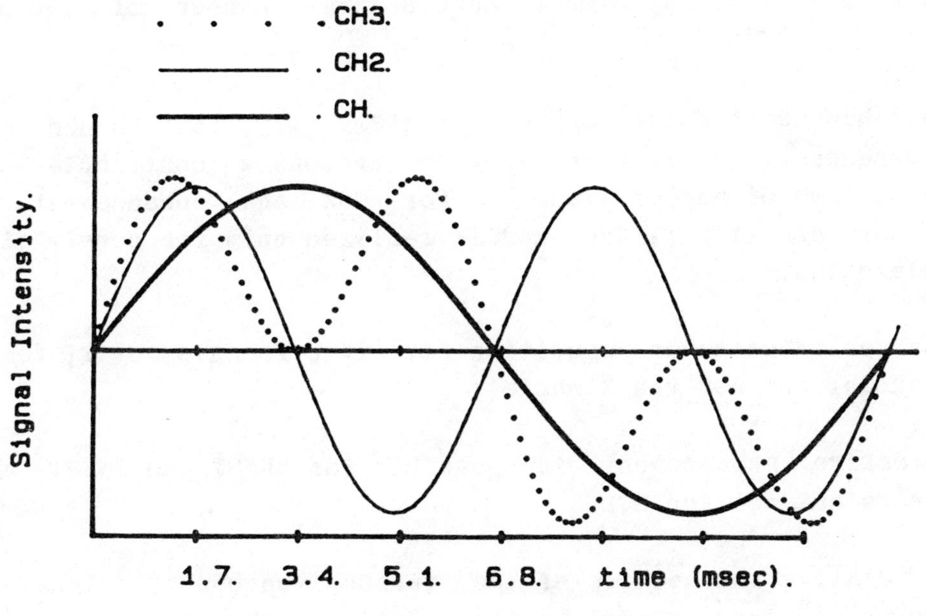

Figure 6.15 The signal intensity as a function of the Δ delay period for CH, CH_2 and CH_3 carbons in an ERNST cross-polarization experiment. τ is fixed at a value of 1/(4J). Note that at an evolution time of about 3.4 msec there is no signal predicted for the CH_2 or the CH_3 carbons, while a maximum enhancement is predicted for the CH carbon signal.

Since the signal enhancement is a function of the value of J_{AX} and the multiplet multiplicity, a number of useful selective enhancements can be obtained if one recalls that the splitting for a CH doublet is J Hz, so that each line is displaced by J/2 Hz from the center of the multiplet. For a CH_2 triplet the outer lines are displaced by the J Hz from the multiplet center. Consequently the phase modulation is twice as fast. As can be seen in Figure 6.15, the maximum signal enhancement for a CH carbon is about 3.4 msec. From Figure 6.15 we also see that for a time $\tau \cong \Delta \cong 1/(4J) \cong 1.7$ msec, we should obtain a nearly full enhancement for CH_2, CH_3, and for CH carbons. For $\Delta \cong 3/(4J) \cong 5.1$ msec, we should see a positive enhancement for the CH and CH_3 carbons and a negative enhancement (that is, an inverted signal) for the CH_2 carbons. Experiments of this sort have become quite common today and for sensitivity enhancement have almost entirely replaced the earlier NOE experiments.

The INEPT and ERNST experiments have a large number of advantages over NOE:

1) An enhancement factor of approximately γ_A/γ_X is obtained and is independent of what relaxation processes contribute to the relaxation of nuclei A and X. For NOE the enhancement is a factor of $(1 + \gamma_A/2\gamma_X)$ and is realized only for purely dipolar relaxation.

2) The pulse sequence repetition rate is a function of T_1 for the A nuclei but not the X nuclei.

3) Selective enhancement is possible for ERNST and INEPT experiments but not for NOE.

4) The relative intensities of the various carbon-13 signals are more nearly proportional to the relative number of the chemically different carbon nuclei present, since for the ERNST experiment the carbon intensities are not dependent on the wide range of NOE and the T_1 values for the different carbons.

An example of the ERNST experiment is given in Figure 6.16 for the camphor molecule whose structure is shown below.

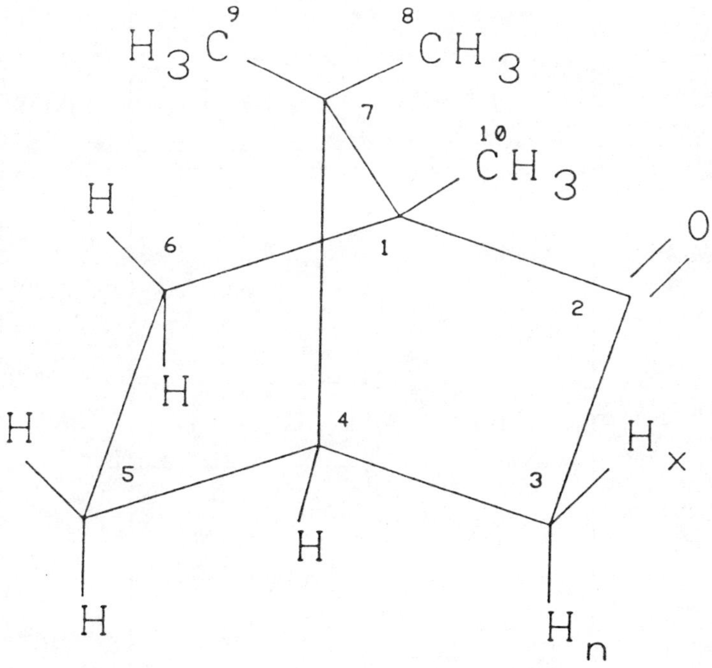

The Camphor Molecule

The proton decoupled carbon-13 spectrum is shown in Figure 6.16c. The C_2 line occurs far downfield from the other carbon resonances and is not shown in the spectrum. The ERNST spectrum shown in Figure 6.16a was obtained by using a dephasing time of about 3.4 msec.; it shows the predicted enhancement for the C_4 (CH) carbon and almost no signal at all for the other carbons. The ERNST spectrum shown in Figure 6.16b shows, as predicted, a positive enhancement for the three methyl carbons (C_8, C_9 and C_{10}), a smaller positive enhancement for the C_4 carbon and a negative enhancement for the CH_2 carbons (C_3, C_5, and C_6). Since only $\pi/2$ and π pulses are used and the pulse sequence repetition rate is fast, the carbon arises solely from polarization transfer from the protons. For this reason, no signal is present for the quaternary carbons in the Ernst experiments.

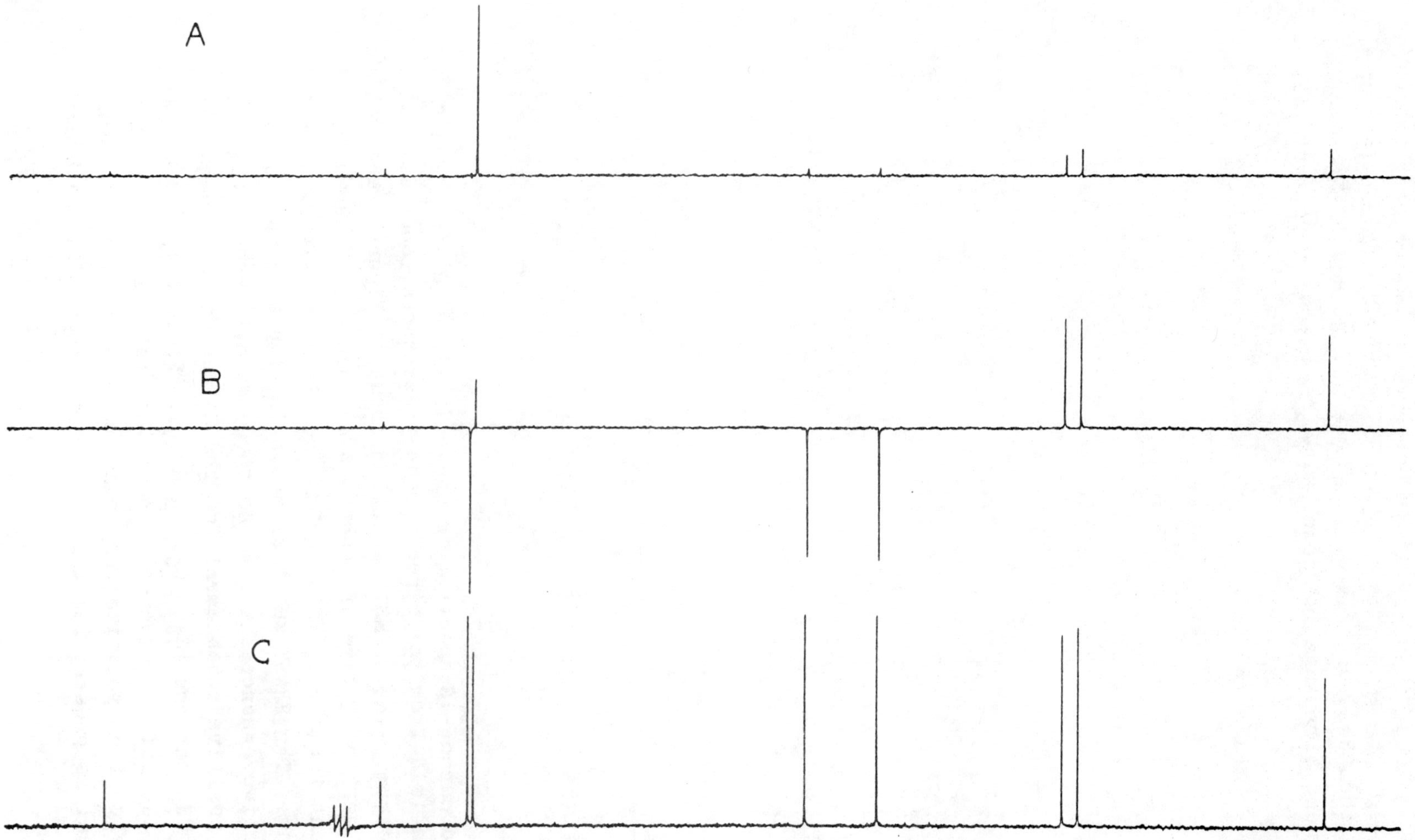

Figure 6.16 (a) The ERNST experiment for camphor run with a delay time of 3.4 msec. In this case only the methine carbon signals are enhanced. (b) With a delay time of 5.1 msec the methine carbon and the methyl carbons have a positive enhancement and the methylene carbons have a negative enhancement, as predicted by equations 6.24a-c. (c) The proton decoupled carbon-13 spectrum of camphor in $CDCl_3$.

An alternative selective polarization enhancement experiment which is not quite as sensitive to the value of the J spin-coupling constant is the DEPT (distortionless enhancement by polarization transfer) experiment (67). Since DEPT depends upon multiple quantum effects, we cannot rely on simple vector diagrams to explain the details. The interested reader is referred to the recent literature (67, 68) for further details.

CHAPTER SEVEN

TWO-DIMENSIONAL NMR SPECTROSCOPY

7.1 Evolution Times in Pulse NMR Spectroscopy.

Most of the more sophisticated NMR experiments involve three distinct time periods: a preparation period, t_0, an evolution period, t_1, and a detection period, t_2; these are shown schematically in Figure 7.1.

```
                                        S(t2)=FID---→-FFT-→→→ F(ω2)

   :      Preparation      :    Evolution       :     Detection      :
   :                       :                    :                    :
___:_______________________:____________________:____________________:__
   :←-------t0------------→:←----------t1------→:←---------t2-------→:
```

Figure 7.1 The three periods of a general NMR experiment: preparation, evolution and detection. The FID signal observed during the detection period is usually represented by $S(t_2)$ and when Fourier transformed yields the frequency domain signal, $F(\omega_2)$.

For example, in an ERNST experiment the preparation period involves a proton relaxation period and a $\pi/2$ pulse on the proton spins. Sometimes the preparation period also requires a number of carbon-13 pulses to assure that $M_0(^{13}C) = 0$ at the end of the preparation period. The evolution period contains both carbon-13 and proton pulses along with evolution times (usually multiples of 1/[4J], where J is the J_{CH} spin-coupling constant). During the detection period for the ERNST experiment the decoupler is turned on and the FID signal is detected (recorded).

After the experiment is completed, the FID signal is Fourier transformed to obtain the frequency domain signal. The ERNST experiment and all of the others discussed up to this point are one-dimensional experiments, since only a single time response is recorded and only a single Fourier transform is required to obtain the desired frequency domain data. The FID signal, as we have seen, is the result of phase modulation arising from precessing nuclear magnetizations with different chemical shifts and spin coupling constants and, consequently, different resonance frequencies. In all of the experiments discussed so far the phase modulation effects of both chemical shifts and spin couplings are usually present during either the t_1 or the t_2 evolution period, or both. By carefully controlling the time periods during which the chemical shift phase modulation is expressed and the periods during which the spin-coupling phase modulation is expressed, we open up an entire new area of NMR experiments, the two-dimensional Fourier transform NMR (2D-FT-NMR) experiments.

In one-dimensional NMR experiments we normally represent the signal analytically as a function of t_2, $S(t_2)$, see Figure 7.1. If we record a large number of such one-dimensional signals as a function of the evolution time t_1, we can represent these different one-dimensional signals by the functions $S_1(t_2)$, $S_2(t_2)$, $S_3(t_2), \ldots, S_n(t_2)$, or, more succinctly, we can represent this collection of FID's as a two-dimensional array $S(t_1, t_2)$. In 2D-FT-NMR spectroscopy, then, we collect an array of one-dimensional signals, which can be Fourier transformed twice to obtain a two-dimensional frequency spectrum, $F(\omega_1, \omega_2)$; thus,

$$S(t_1, t_2) \text{ ---2 x FFT} \rightarrow\rightarrow\rightarrow \quad F(\omega_1, \omega_2). \qquad (7.1)$$

In 2D-FT-NMR experiments the structure of the array, $S(t_1, t_2)$, depends primarily on the following parameters:

1) the chemical shifts and coupling constants present and the periods during which they are expressed

2) the evolution periods (t_1) used

3) the pulses which are applied and the times during which the decoupler is operating.

In most of the experiments which follow we arrange the experimental conditions such that only one parameter (e.g., spin coupling) is expressed during the evolution time and a second parameter (e.g., chemical shift) is expressed during the detection period. This allows one to have a final two-dimensional frequency spectrum in which one parameter (e.g., chemical shift) is expressed along one of the frequency dimensions and the second parameter (e.g., spin coupling) is expressed along the second frequency dimension. Refocusing pulses are used to eliminate the expression of chemical shifts (or in some cases spin-coupling constants), and the decoupler can be used to suppress the expression of spin coupling during the various time periods.

In order to make these concepts clearer, let us consider a number of simple experiments. In particular, let us consider the carbon-13 and proton spectra of $^{13}CHCl_3$. (Although we will use carbon and proton spins, we could equally well use two general spins, A and X, both with spin 1/2, for the following discussion.) The first experiment we consider is shown in Figure 7.2.

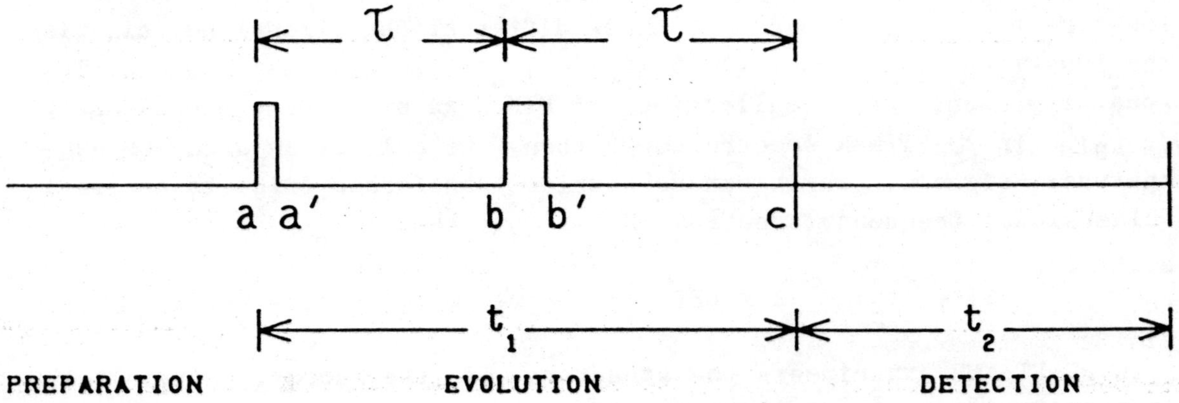

Figure 7.2a The three time periods for a typical NMR experiment consist of the preparation period, t_0, the evolution period, t_1, and the detection period, t_2. The time τ between the $\pi/2$ and the π pulses is one-half the evolution time, t_1.

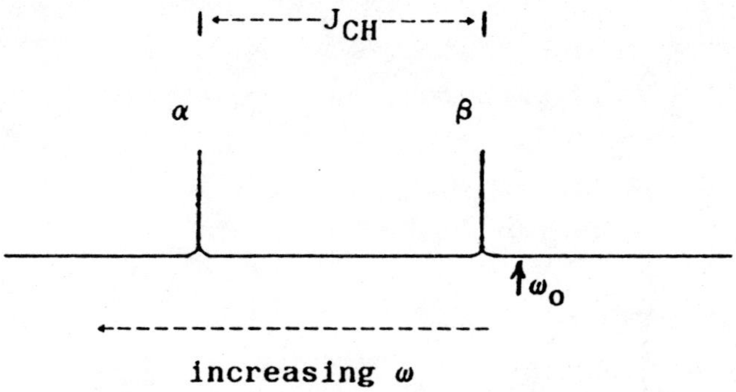

Figure 7.2b The carbon-13 NMR spectrum of $^{13}CHCl_3$. The spectrometer reference frequency (or the frequency of the rotating frame), ω_0, is just slightly less than the frequency of the β carbon line in the spectrum.

The spectrometer reference frequency, ω_0, is slightly less than the resonance frequency for the carbon nucleus spin coupled to the proton in the β spin state. In a time $\tau = t_1/2 = 1/(4J)$, there will be a dephasing of the carbon magnetization due to both chemical shift effects and spin-coupling interactions. In the present case, in a time τ, the α carbons (i.e., those carbons which are spin-coupled to a proton in the α spin state) will be 90° out of phase with the β carbons. A carbon π_y pulse initiates the refocusing of both the chemical shift dephasing (or precession) and the spin-coupling dephasing (see b and c in Figure 7.3). At time t_1 (c in Figure 7.3), both the chemical shift and the spin-coupling dephasing are refocused.

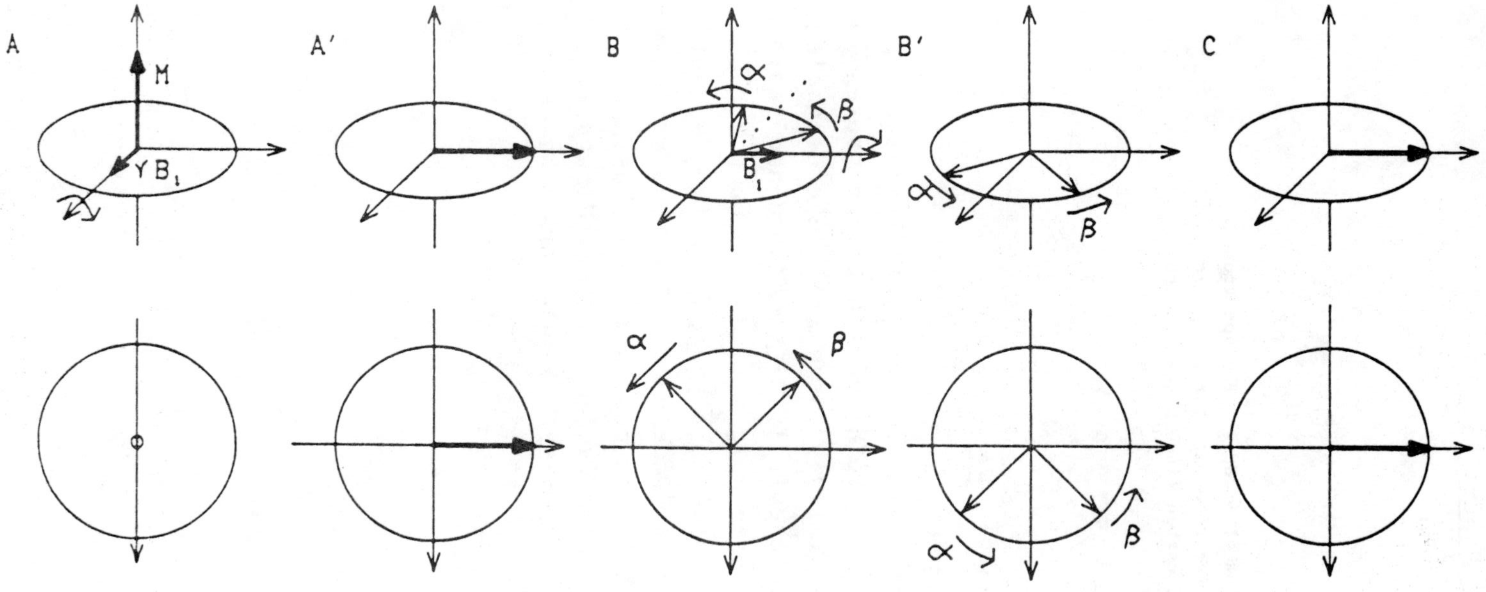

Figure 7.3 A carbon $(\pi/2)_x$ rotates the magnetization into the xy plane, where it begins to dephase due to chemical shift and spin-coupling interactions. After an evolution time $\tau = (t_1)/2 = 1/(4J)$, the carbon α and β magnetizations are 90^0 out of phase with one another. A carbon π_y pulse initiates the refocusing of both the chemical shift and the spin coupling dephasing. After a further evolution time $\tau = 1/(4J)$, the magnetizations are completely refocused.

7.2 J-Resolved Two Dimensional NMR.

Consider next an experiment in which we refocus the chemical shift dephasing but not the spin-coupling dephasing during the evolution time. The pulse sequence used here is shown in Figure 7.4

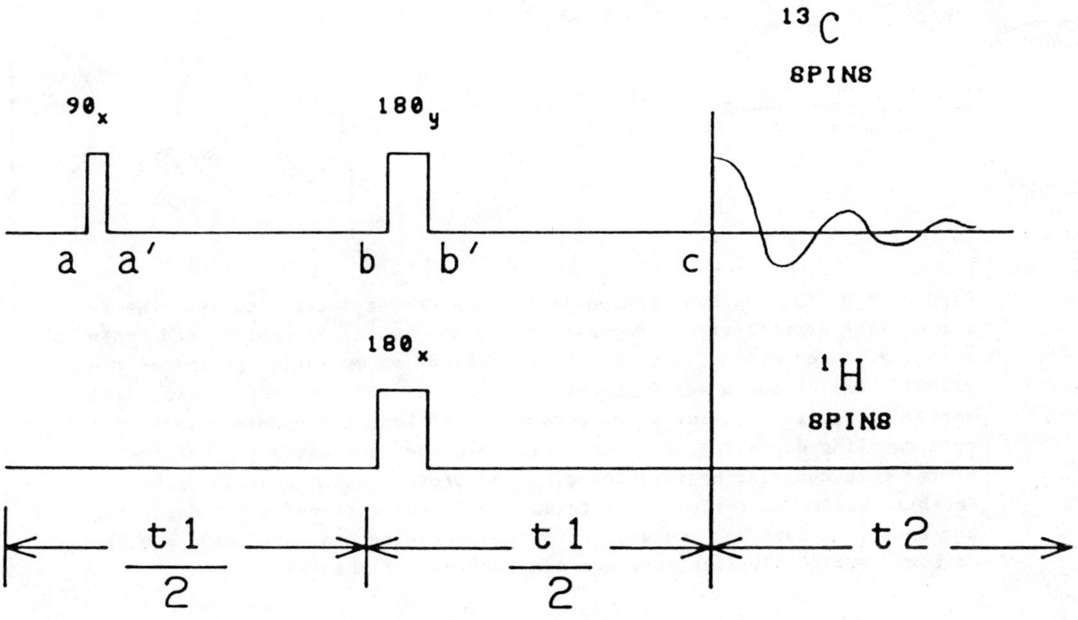

Figure 7.4 In this pulse sequence the carbon π pulse at the midpoint of the evolution period initiates the refocusing of the effect of the carbon chemical shift. The proton π pulse suppresses the refocusing of the spin coupling by interchanging the proton spin states.

The dephasing and refocusing is shown in the vector diagrams in Figure 7.5.

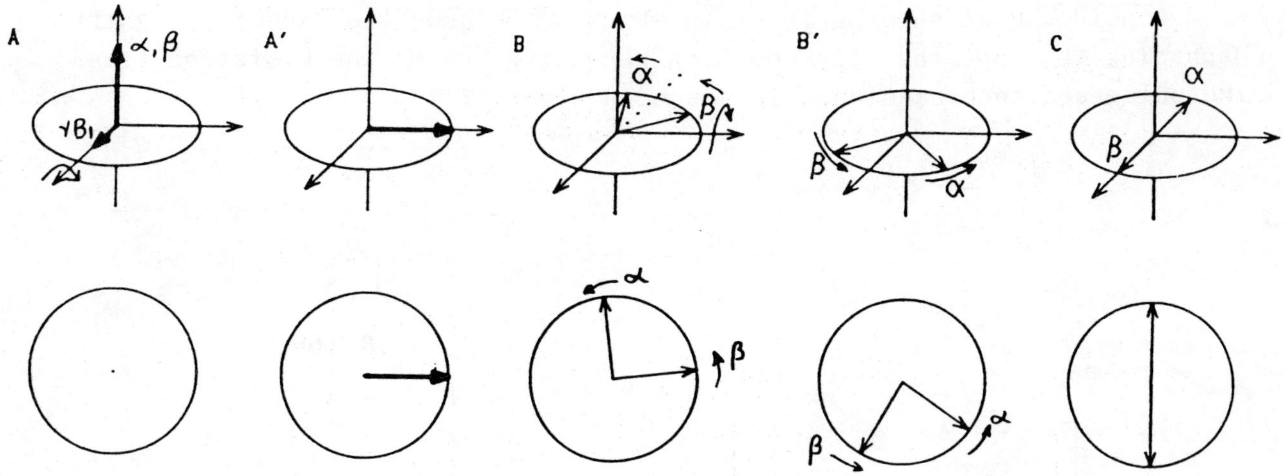

Figure 7.5 The carbon $\pi/2$ pulse rotates the magnetization into the xy plane. The magnetizations dephase due to the chemical shift and spin coupling interactions. At point b, after an evolution period of $\tau = 1/(4J)$, the carbon α and β magnetizations are 90° out of phase. The carbon π pulse initiates the refocusing of both the chemical shift and spin-coupling dephasing, but the proton π pulse suppresses the refocusing of the spin coupling by interchanging the proton spin states. After a further evolution period, $\tau = 1/(4J)$ (c), the chemical shift dephasing has been refocused but the spin-coupling dephasing has continued, and the carbon α and β magnetizations are now 180° out of phase.

Here the π_x pulse on the proton spins interchanges their spin states; consequently, the spin-coupling effects cause the carbon magnetizations to continue to dephase at point b' rather than refocus. However, the chemical shift dephasing is refocused. The above two experiments are similar to a Carr-Purcell experiment or to the first part of an INEPT or ERNST experiment. The following experiment is quite different because we add the possibility of switching on the proton decoupler during part of the evolution time and during the acquisition. The effect of turning on the decoupler is to suppress the dephasing due to spin coupling. The pulse sequence for executing this experiment is shown in Figure 7.6, and the magnetization vector diagram is shown in Figure 7.7

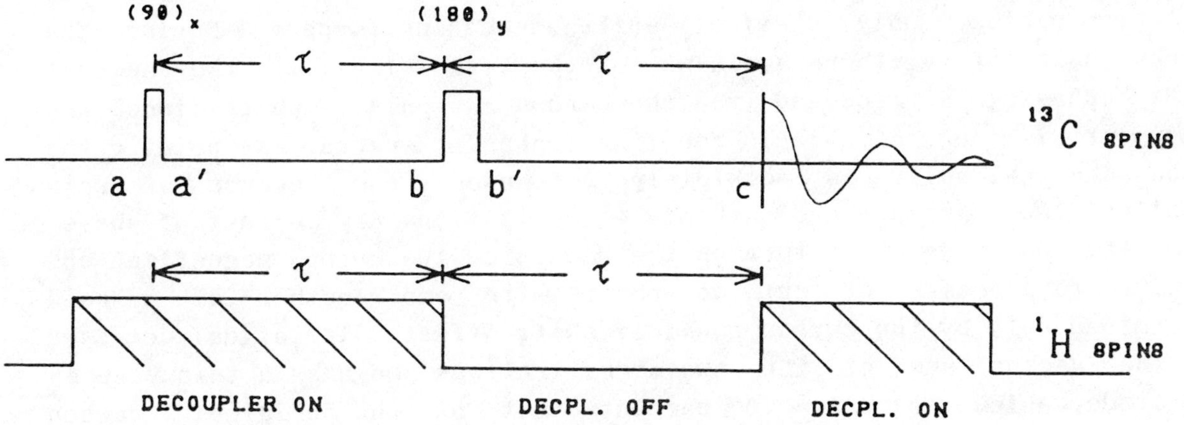

Figure 7.6 A pulsed NMR experiment in which spin-coupling effects are suppressed during the first part of the evolution period and during the acquisition period.

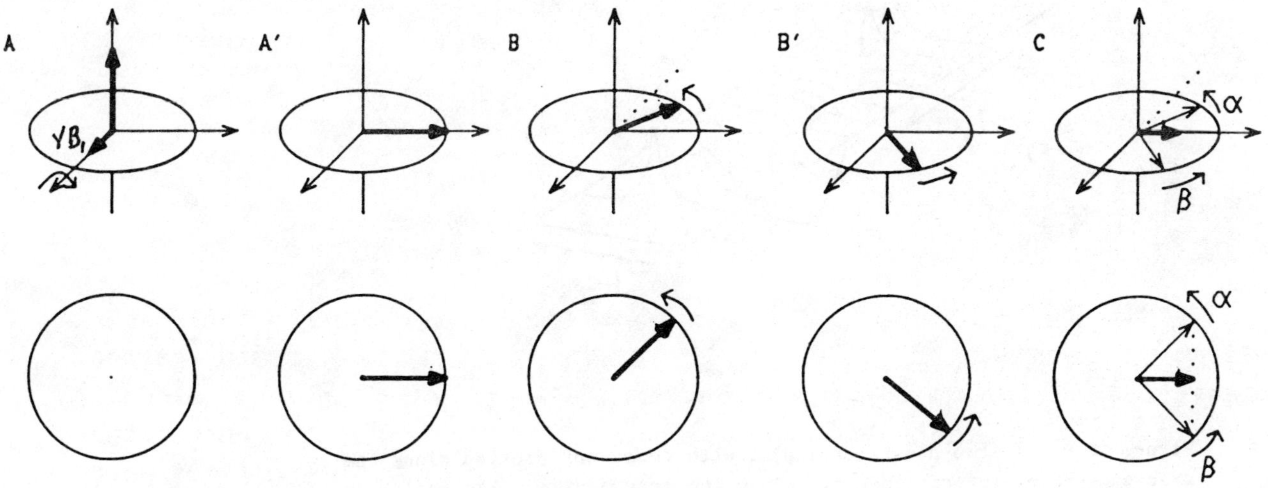

Figure 7.7 During the first part of the evolution period only chemical shift dephasing is expressed (during the time between a' and b). From b' to c the chemical shift dephasing is refocused and spin-coupling dephasing takes place. During the detection period futher dephasing due to spin coupling is suppressed.

Since the proton decoupler is on during the first half of the t_1 evolution period, only chemical shift dephasing occurs. During the second half of t_1 (here again we let $\tau = t_1/2 = 1/[4J]$), the chemical shift dephasing is refocused (via the carbon π_y pulse), but since the decoupler is now off, spin-coupling dephasing occurs. At point c the carbon chemical shift is completely refocused, and because of spin coupling the carbon α and carbon β magnetizations are 90^0 out of phase. If at this point we again turn on the decoupler the carbon magnetizations continue to precess 90^0 out of phase with one another at a rate determined only by the carbon chemical shift offset. The signal detected is the vector sum of the two magnetizations and has in this case an amplitude which is only 70 percent that of the original carbon magnetization. If we change the evolution time such that $\tau = 1/(2J)$, the final proton-decoupled carbon magnetizations will be 180^0 out of phase and the resulting carbon signal will be zero. This is summarized graphically in Figure 7.8.

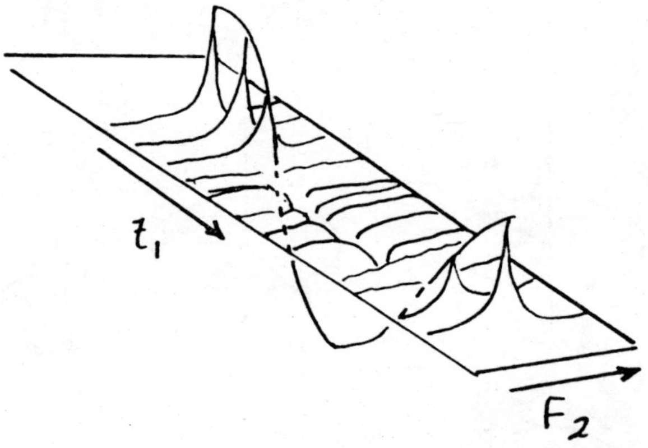

Figure 7.8 A two-dimensional plot with frequency plotted along the F_2 axis and the evolution time, t_1, along the second axis. The amplitude of the signal is a function of the spin-coupling dephasing which took place during the second part of the evolution period (see Figure 7.7).

In Figure 7.8, then, we see a two-dimensional plot of the results where the transformed frequency spectra are given along the F_2 axis and the evolution time is plotted along the t_1 axis. If we look along the t_1 axis we see that the signal amplitude varies sinusoidally as a function of the evolution time, t_1. If we Fourier transform this t_1 waveform we obtain a frequency equal to J/2. This gives peaks at ± J/2. The two-dimensional frequency plot is given in Figure 7.9.

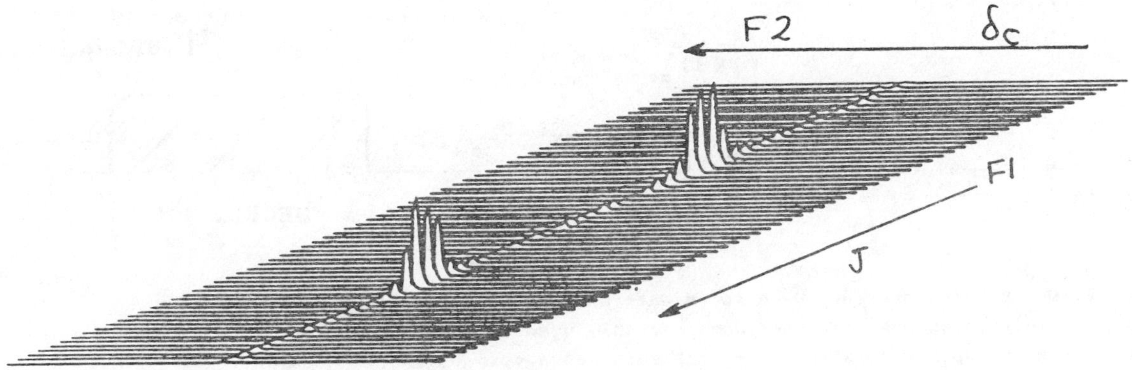

Figure 7.9 A two-dimensional transform of the experiment shown in Figure 7.7. Chemical shift dephasing is refocused in the evolution period t_1 and spin-coupling dephasing is suppressed during the detection period t_2. Because t_1 contains spin-coupling information only and t_2 contains chemical shift information only, the F_1 axis shows spin-coupling constants and the F_2 axis shows chemical information.

Since the source of the amplitude modulation along t_1 is the expression of the spin coupling, the F_1 frequency gives the spin-coupling constant and the F_2 frequency gives the chemical shift value.

The experiments above are useful for understanding the basic concepts of two-dimensional NMR but are not very useful for most NMR experiments. One of the simplest and most useful 2D-NMR experiments is the J-resolved 2D-FT-NMR experiment. The pulse sequence used for the heteronuclear version of this experiment is shown in Figure 7.10.

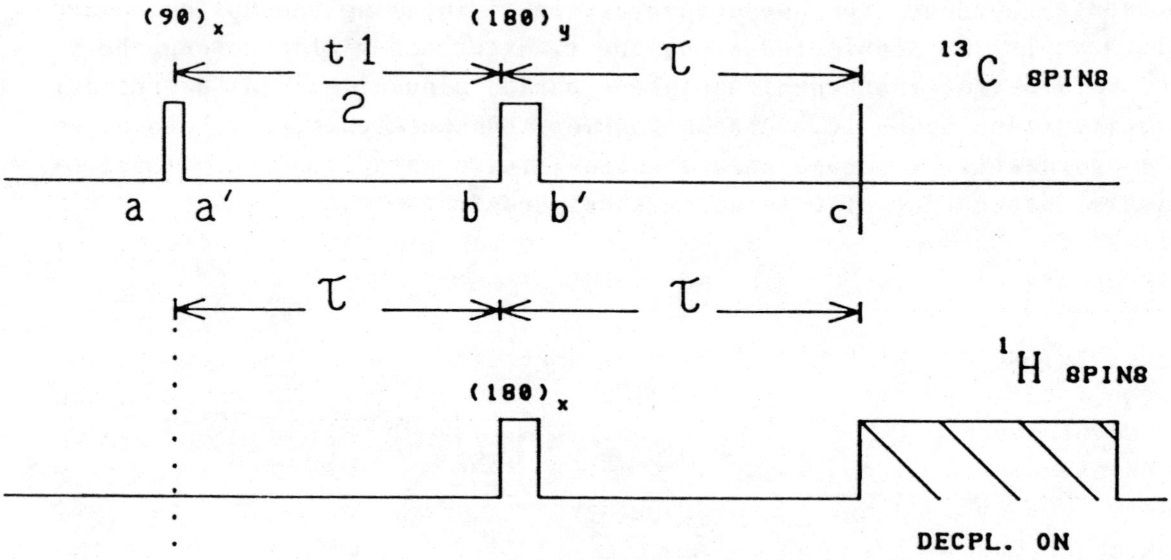

Figure 7.10 The pulse sequence used for a heteronuclear J-resolved 2D-NMR spectrum. This sequence is the one commonly used. This example is, again, for a two-spin AX system where each spin has a spin of 1/2.

Figure 7.11 shows the evolution of the carbon magnetization, again for the two spin system, $^{13}CHCl_3$.

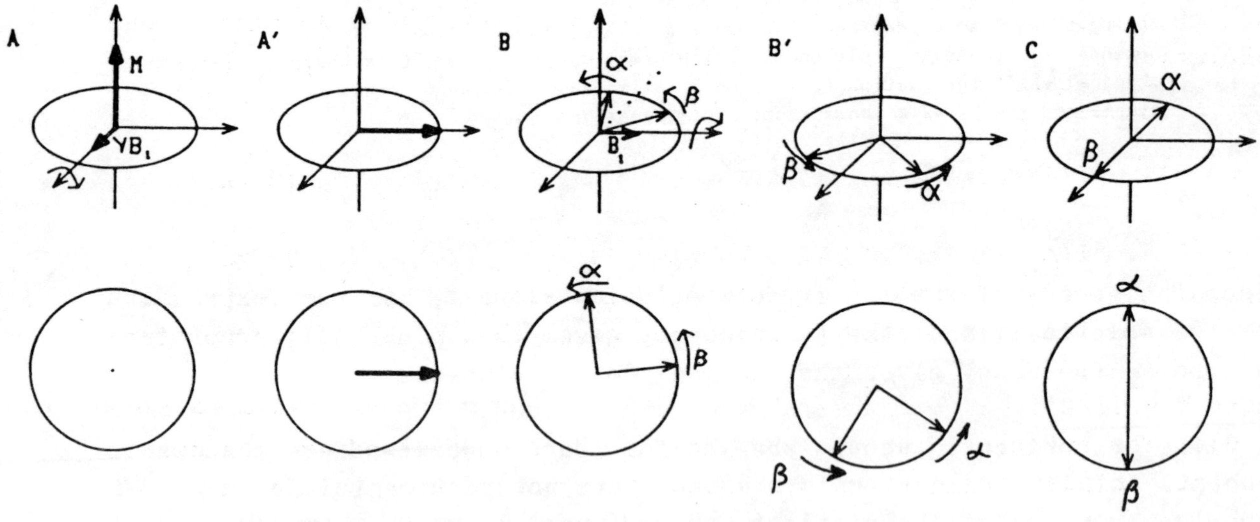

Figure 7.11 The vector diagram for the evolution of the carbon magnetizations for the pulsed NMR experiment shown in Figure 7.10.

We see in this relatively simple experiment, first done by Ernst and others (69-73), that only spin-coupling phase modulation is expressed during the evolution period t_1 (the chemical shift phase modulation is refocused). During the detection period, t_2, spin-coupling phase modulation is suppressed by turning on the decoupler, consequently, during this period only chemical shift phase modulation is expressed. This summarizes the basic concept of 2D-FT-NMR spectroscopy. Whenever we have a systematic variation of a signal (i.e., a phase modulation) due to a single property of a spin system (e.g., chemical shift or spin coupling) during a given evolution time, we can separate the effects of that property using two-dimensional spectroscopy.

Up to this point we have considered only two spins, carbon and hydrogen (or, in general, X [carbon] and A [hydrogen]). The concepts and ideas developed for two spins can readily be generalized to more complex spin systems. For the three-spin system, A_2X, the X spectrum is a triplet. The outer lines of the triplet are phase modulated in the same way as an AX doublet, but at a frequency of +J and -J with respect to the central peak. The central peak is not phase modulated at all due to spin-coupling effects. In other words, the outer pair of lines for the triplet behave like an AX doublet, but with twice the phase modulation frequency (recall that the AX doublet is modulated at a rate $\pm$ J/2). An A_3X exhibits a quartet spectrum similar to a doublet of the AX system, but, of course, in a different way. The modulation frequency for the inner pair of lines in the A_3X quartet is $\pm$ J/2, exactly the same as for the AX doublet. The modulation frequency for the outer pair of lines is $\pm$ 3J/2. Another way to express this is to write the expressions for the amplitude modulation, I, of the receiver signal due to the phase modulation of the AX, A_2X, and A_3X spin couplings. These are as follows:

$$I(AX) = M_o\cos(\pi\tau J), \qquad (7.2a)$$

$$I(A_2X) = (1/2)M_o[1 + \cos(2\pi\tau J)], \qquad (7.2b)$$

$$I(A_3X) = (1/4)M_o[3\cos(\pi\tau J) + \cos(3\pi\tau J)], \qquad (7.2c)$$

where $\tau = (1/2)t_1$ is the time between the $\pi/2$ and π pulses (see Figure 7.10). The vector diagrams showing the phase modulation for a singlet, doublet, triplet and quartet are shown in Figure 7.12.

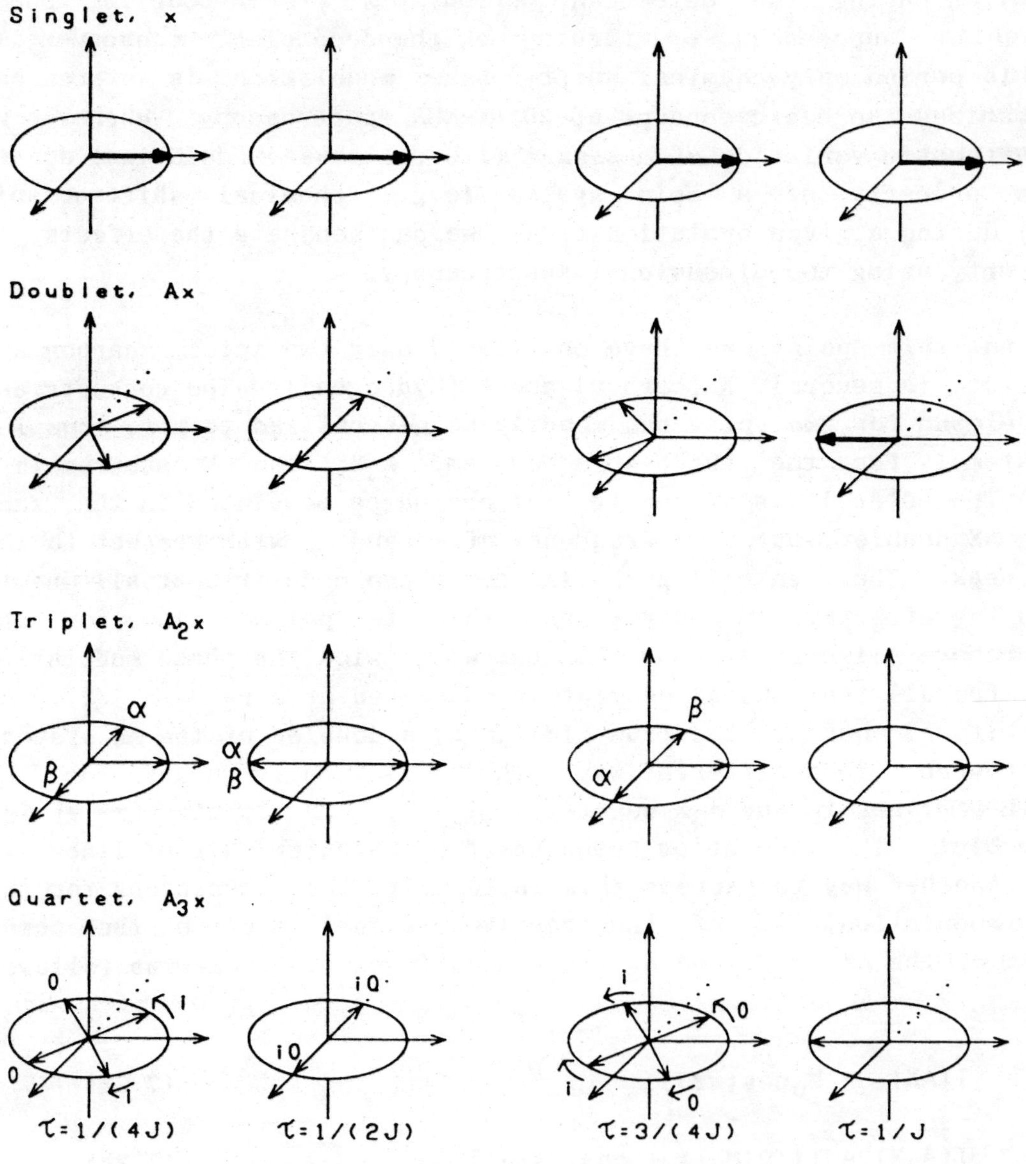

Figure 7.12 This set of vector diagrams shows the evolution of (a) an X nucleus with no directly coupled A nuclei (b) the dephasing of the X magnetization for an AX system, (c) the dephasing for the X magnetization for an A_2X system, and (d) the dephasing for the lines in an A_3X quartet. The diagrams in the first column are for an evolution time τ = 1/(4J). the second column diagrams are for τ = 1/(2J). the third column diagrams are for τ = 3/(4J) and the last column diagrams are for a dephasing time. τ = 1/J.

By making use of this ability to express the phase modulation of the chemical shift, the spin coupling, or both, at various times, a number of very useful two-dimensional experiments can be performed. In Table 7.1 we summarize some of the more common and useful experiments, and indicate the information which may be obtained with them.

Table 7.1

Experiment	F1	F2	Information
Heteronuclear J-resolved	J_{AX}	δ_X	Heteronuclear coupling constants
Homonuclear 2D-J	$J_{A,A}$	δ_A *	Homonuclear J & δ
Correlated Heteronuclear 2D shift spectroscopy (CHESY) or HETCOR	δ_A	δ_X	Correlation of δ_X & δ_A
Homonuclear correlated Spectroscopy (COSY)	δ_A	δ_A	Identification of all scalar coupling interactions
2D-NOE	δ_H $J_{H,H}$	δ_H $J_{H,H}$	Spatial proximity & 3D molecular structure
2D-INADEQUATE	$\delta_A + \delta_X$	δ_X	Heteronuclear connectivities

* After "tilting"

7.3 Heteronuclear Correlated Spectroscopy.

As a further example, let us consider what happens in a correlated heteronuclear spectroscopy (HETCOR or CHESY) experiment (73-75). In Figure 7.13 we show the proton spectrum for $^{13}CHCl_3$, and in Figure 7.14 we show the pulse sequence used.

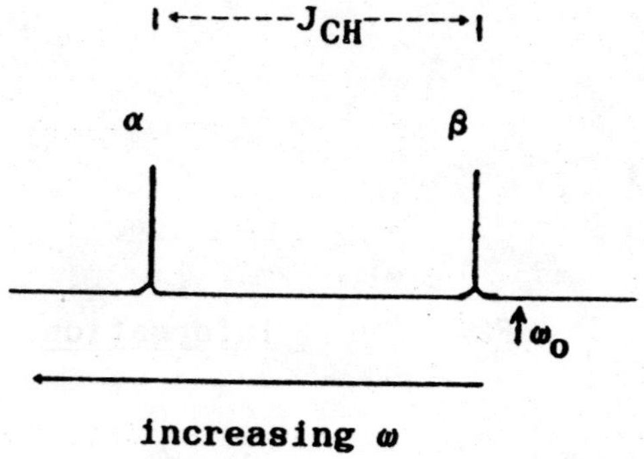

Figure 7.13

Figure 7.13 The proton NMR spectrum for the molecule $^{13}CHCl_3$.

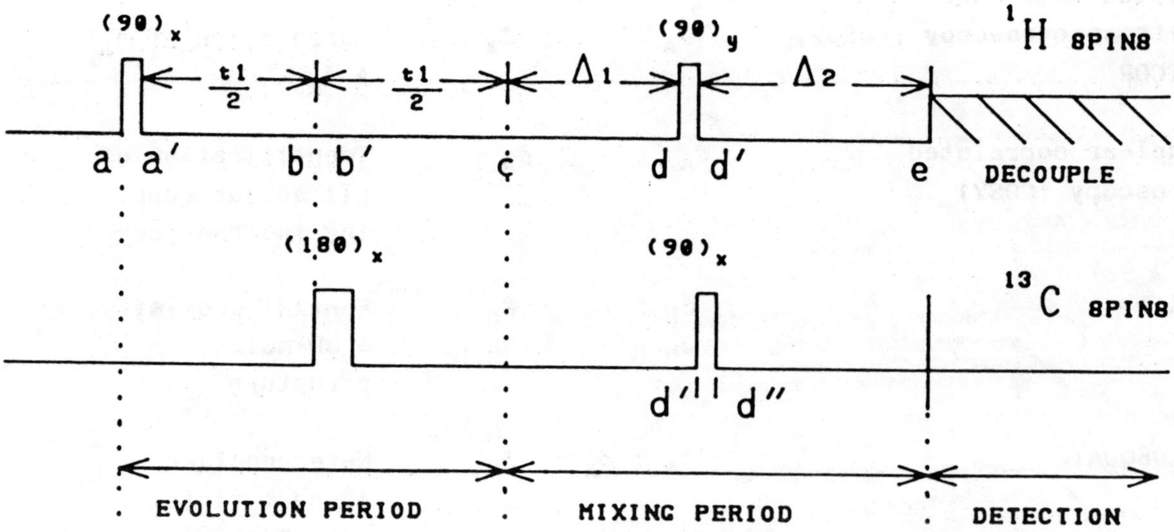

Figure 7.14 The pulse sequence used for a correlated heteronuclear spectroscopy (HETCOR or CHESY) spectrum. The τ times are varied to obtain the intervals in the F1 dimension. In the experiment additional mixing times, Δ1 and Δ2 are needed. In most cases, Δ1 = 1/(2J).

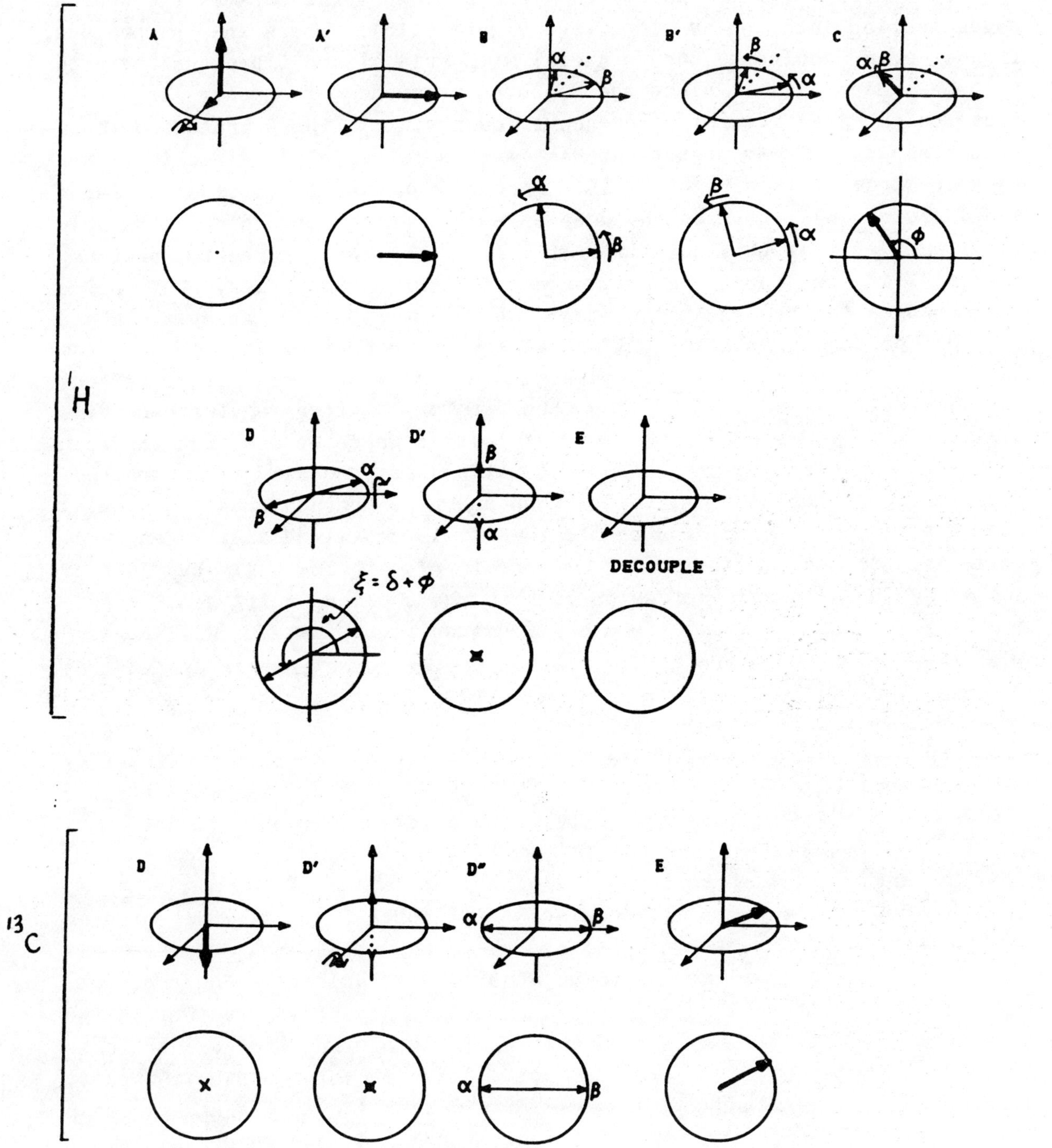

Figure 7.15. The top four rows show the evolution of the A (proton) spins in a CHESY experiment. The proton magnetization is rotated into the xy plane (a'), where it undergoes chemical shift and spin-coupling dephasing. See the text for a detailed explanation of dephasing and refocusing. The last two rows show the dephasing for the carbon magnetization.

The first $(\pi/2)_x$ pulse on the proton spins rotates the proton magnetization into the xy plane (see Figure 7.15), where the α protons (those spin coupled to carbon nuclei with spin α) and β protons begin to dephase. Here again, since the β proton resonance frequency is quite close to the spectrometer reference frequency, ω_o, the β protons dephase rather slowly. The α protons dephase much more rapidly, since they are further from ω_o (see Figure 7.13). After dephasing for a time τ, the α and β protons will have dephased by an amount which depends on the spin coupling, J_{CH}. At this point (point b in the vector diagram for protons, Figure 7.15) we apply a π_x pulse to the carbon spins. This pulse interchanges the carbon spin states, which interchanges the spin labels on the proton magnetizations and causes the dephasing that arises from the spin coupling to begin refocusing. This does not, of course, refocus the chemical shift dephasing for the protons. After a further time evolution of τ seconds we are at point c where we see that the spin coupling has refocused but the magnetization has precessed (or dephased) due to the proton chemical shift by an amount ϕ degrees from the y axis. Information about the proton chemical shift is contained in this angle ϕ. If we wait for a further time interval $\Delta 1$, there will be further dephasing due to the expression of both the chemical shift and the spin coupling. If $\Delta 1 = 1/(2J)$, the α and β proton magnetizations will be 180° out of phase with one another. A $(\pi/2)_y$ proton pulse at this point (d) rotates the antiparallel proton magnetizations along the z axis (d'). This is reminiscent of the ERNST and INEPT experiments; not surprisingly, then, we get a polarization transfer to the carbon spins. The amount of polarization transferred depends on the angle ϕ, since the $(\pi/2)_y$ pulse affects only the x components of the α and β proton magnetizations.

The angle which these antiparallel magnetizations make with the y axis depends on the amount of chemical shift dephasing, ϕ, which occured during the time 2τ and the amount, δ, which occurred during the time $\Delta 1$. Now the time $\Delta 1$ is fixed by the value of the AX spin-coupling constant, J_{AX}, so the amplitude <u>modulation</u> is a function only of the evolution time t_1, that is, it depends only on the magnitude of proton chemical shift precession angle, ϕ. If we let $\xi = \phi + \delta$, then the proton chemical shift information is contained in the angle ξ. For $\xi = (2n + 1)(\pi/2)$, where n is an integer, the proton magnetizations at point d, and hence the amount of cross polarization, will be a maximum. For $\xi = n\pi$ there will be no longitudinal proton magnetization and, consequently, no cross-polarization of the carbon spins. The result is that the magnitude of the carbon magnetization measured during the detection period depends directly on the angle ϕ, which means that the magnitude of a particular carbon signal depends upon the chemical shift of the proton to which it is spin coupled. In other words, variation of the evolution time t_1 causes a variation or modulation of the observed signal. This effect on the carbon signal is propagated via the spin coupling of the directly bonded protons, but it is not a function of the magnitude of J_{CH}; it depends only on the value of the proton chemical shift, δ_H (see Figure 7.13).

The pulse on the carbon spins rotates the antiparallel α and β carbon spins into the xy plane, and the Δ2 (=1/[2J]) time interval allows the carbon magnetizations to refocus (as in the ERNST experiment). During the detection period the proton decoupler is turned on, so only the chemical shift of the carbon magnetization is expressed during this t_2 interval. The overall result is that proton chemical shift dephasing, δ_H, is expressed during the evolution period t_1 and only carbon chemical shift, δ_C, dephasing is expressed during the detection period. A two-dimensional Fourier transform of $S(t_1, t_2)$ yields a 2D-FT NMR frequency spectrum $F(\omega_1, \omega_2)$ with the proton chemical shifts along the F_1 axis and the carbon chemical shifts along the F_2 axis. This result for $^{13}CHCl_3$ is shown as a contour plot in Figure 7.16.

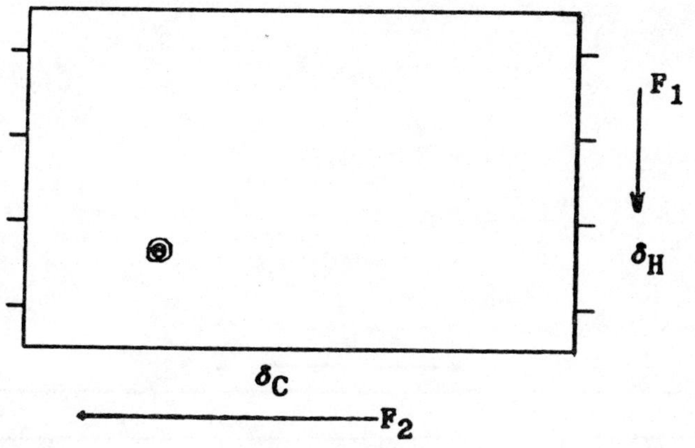

Figure 7.16 The two-dimensional CHESY spectrum for a sample of $^{13}CHCl_3$. The F1 dimension shows the chemical shift of the proton and the F2 dimension shows the chemical shift of the carbon.

Here, in contrast to the SPI experiment where carbon-proton correlations are worked out on a selective one-carbon-at-a-time basis, the heteronuclear experiment shows simultaneously all of the carbon-proton correlations. The results for methyl vinyl ketone are shown in Figure 7.17. This magnificent experiment provides in one fell swoop the following: carbon sensitivity enhancement, elimination of the problem of spectral line overlap, and simultaneous correlation of all carbon and proton resonances in an unambiguous fashion. In addition, it does all of these things much faster than any other method. This experiment is especially useful for complex molecules and/or for situations where there are several scalar coupling constants between the protons and carbons (or other proton-coupled X nuclei) whose values are almost the same. For more examples and details see references 65 and 75.

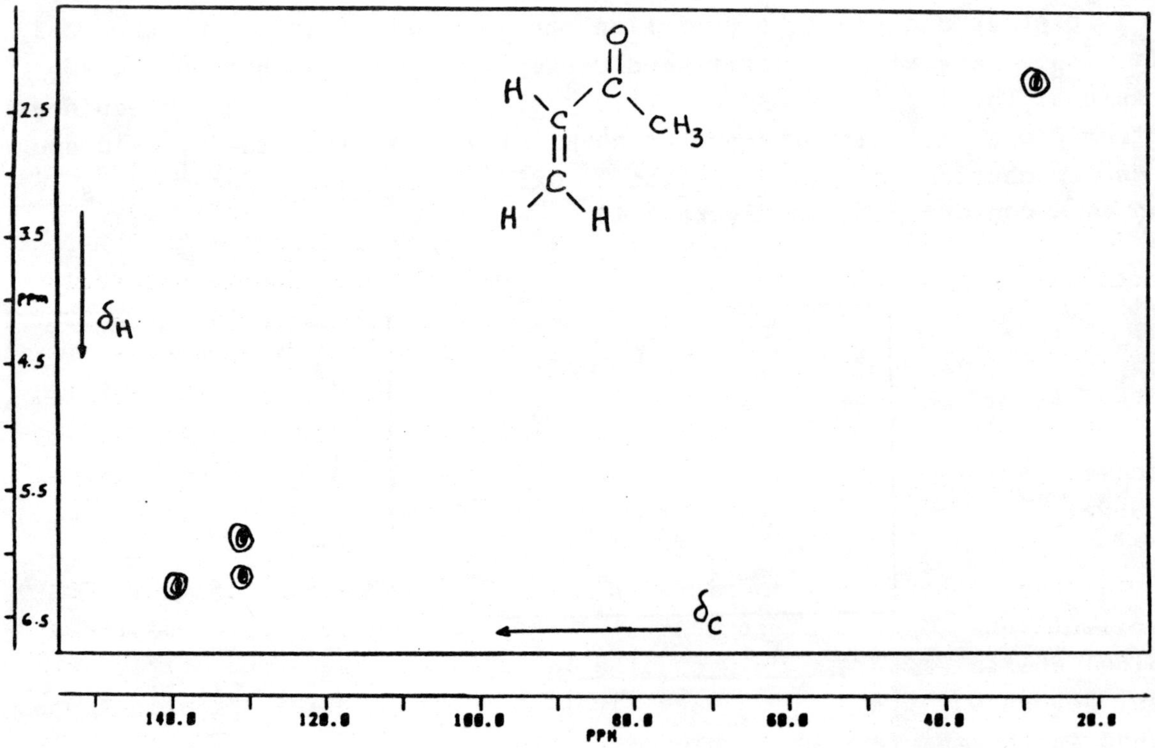

Figure 7.17 The 2D-CHESY spectrum for a sample of methyl vinyl ketone, $H_3C\text{-}C(O)\text{-}CH{=}CH_2$. There are two protons correlated with the $=CH_2$ carbon, since these two protons are chemically not equivalent.

Consider now the 2-D ^{13}C-^{1}H correlation spectrum of methyl vinyl ketone, H_3C^a-$C^b(O)$-C^cH=C^dH_2. The methyl carbon, C^a, at about 20 ppm chemical shift, is correlated with the methyl protons at about 2.5 ppm. The C^c at 140 ppm is correlated with the proton at about 6.4 ppm. Since the two protons bonded to the C^d carbon are not chemically equivalent, we see two contours at a carbon chemical shift of 130 ppm, one for each of the two protons (at about 5.8 and 6.3 ppm). The ketone carbon, C^c, has no directly bonded proton and does not show up on the heteronuclear correlated 2D-FT-NMR spectrum.

7.4 Homonuclear Correlated Spectroscopy.

In homonuclear correlated 2-D NMR spectroscopy (COSY) both the F1 and F2 axes contain chemical shift information. This experiment is very useful, since it allows one to obtain information about the connectivities of the various nuclei. Most COSY experiments are variations of the ones first done by Jeener (69) and Ernst (70). The COSY pulse sequence is shown in Figure 7.18. The experiment depends upon the coherent transfer of magnetization via spin-coupling. A second version of the COSY experiment depends upon the incoherent transfer of magnetization, via the nuclear Overhauser interaction (NOESY) or chemical exchange.

In the coherent magnetization transfer variation of the COSY experiment one observes correlations between nuclei with different chemical shifts which are spin coupled to one another. The first $\pi/2$ pulse generates transverse magnetization. The various magnetizations precess in the xy plane at a rate determined by their chemical shift values and their spin coupling to other, chemically different, nuclei. For two nuclei A and X which are not spin coupled ($J_{AX} = 0$), the second $\pi/2$ pulse induces a time dependent (t_1, see Figure 7.18) modulation of the magnetization which depends only on ω_A for nucleus A or only on ω_X for nucleus X. The two-dimensional spectrum for uncoupled nuclei, therefore, contains peaks only at the positons for which F1 = F2, that is, it contains peaks only along the diagonal.

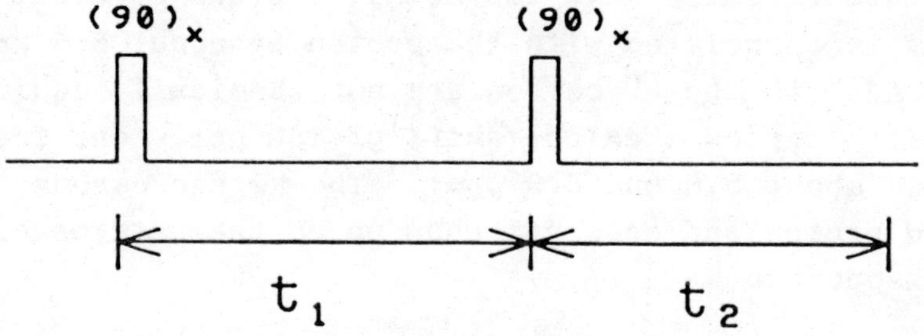

Figure 7.18 The pulse sequence for a COSY NMR experiment. The first $\pi/2$ pulse generates the transverse magnetization in the xy plane. The second $\pi/2$ pulse is, effectively, a mixing pulse. It causes an exchange of magnetization between spin coupled nuclei.

If $J_{AX} \neq 0$, then the precession rate of the A magnetization in the xy plane depends upon both ω_A and J_{AX}; a similar situation holds for the X magnetization. This will produce off diagonal peaks with F1 and F2 coordinates of ω_A, ω_X and ω_X, ω_A. An example, the ^{11}B COSY spectrum of 2-$(CH_3)B_5H_8$ is shown in Figure 7.19. The one-dimensional boron-11 NMR spectrum is shown along the upper hirizontal axis. In the COSY spectrum there are cross peaks between boron nuclei 1, 2, 3, and 5 and between 1, 3, 4, and 5. There are no cross peaks between nuclei 2 and 4, since there is no spin coupling between nuclei 2 and 4. It is interesting to note that, if the experiment is done carefully, cross peaks are observed for bridge bonded, B-H-B, boron nuclei (76).

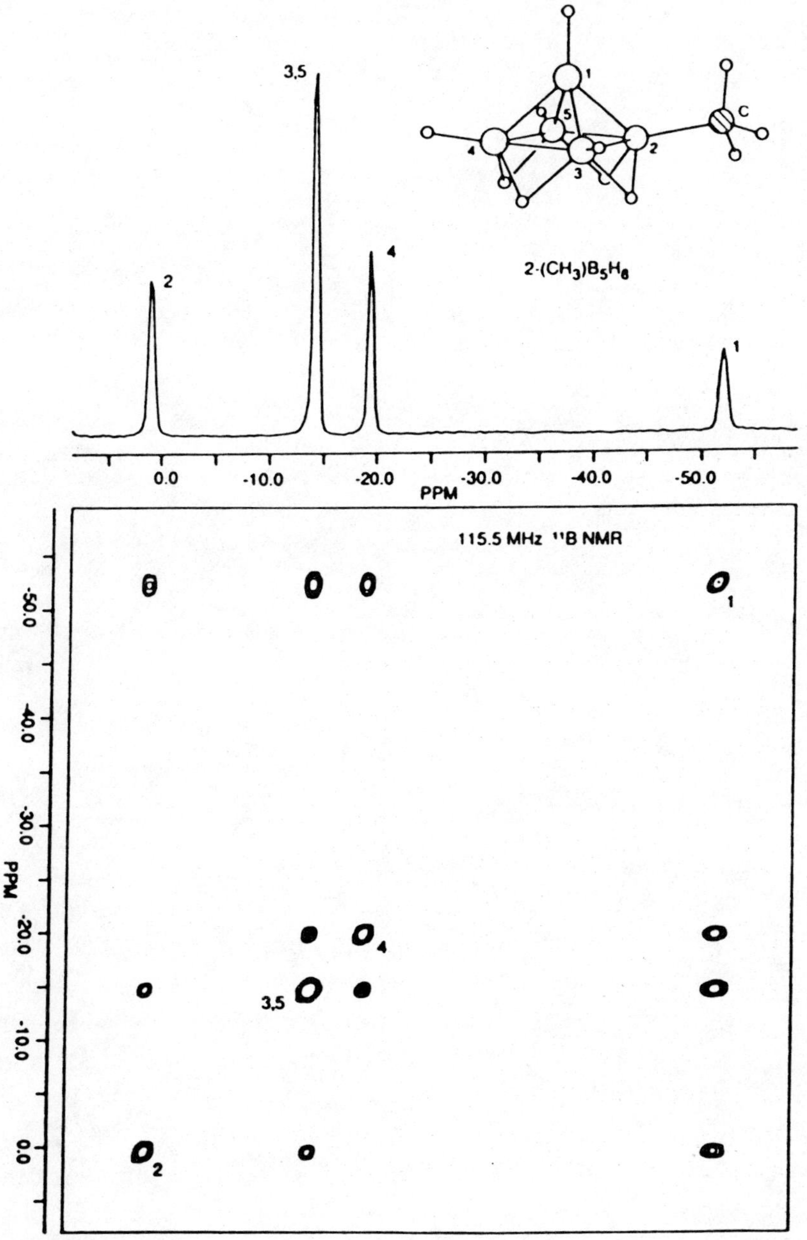

Figure 7.19 The ^{11}B COSY spectrum of 2-$(CH_3)B_5H_8$. There are ^{11}B cross peaks between nuclei 1, 2, 3, and 5 and between 1, 3, 4, and 5. There are no cross peaks between nuclei 2 and 4, because there is no spin coupling between these nuclei. The one-dimensional ^{11}B NMR spectrum appears along the diagonal.

Figure 10. The ^{1}H COSY spectrum of [illegible]. There are [illegible] cross peaks between [illegible] and [illegible] between [illegible] are [illegible] between [illegible] and 4, [illegible] coupling between these nuclei. The one-dimensional ^{1}H NMR spectrum appears along the diagonal.

APPENDIX A

A Brief Review of Vectors

A vector has both magnitude and direction; it is denoted in this book by boldface type. A vector is usually written in terms of its components along three mutually orthogonal directions. For example, the vector **V** is written as

$$\mathbf{V} = \mathbf{i}V_x + \mathbf{j}V_y + \mathbf{k}V_z \ , \qquad \text{(A-1)}$$

where **i**, **j**, and **k** are vector of unit length, directed along the x, y, and z axes, respectively. The length, or magnitude, of a vector is given by

$$|\mathbf{V}| = (V_x{}^2 + V_y{}^2 + V_z{}^2)^{1/2}. \qquad \text{(A-2)}$$

Usually, if there is no ambiguity, we shall use V, rather than |**V**|, to denote the magnitude of the vector **V**.

The scalar, or dot, product of two vectors, **R** and **S**, oriented at an angle θ with respect to one another, is given by

$$\mathbf{R} \cdot \mathbf{S} = |\mathbf{R}| \ |\mathbf{S}| \cos\theta \ . \qquad \text{(A-3)}$$

It is straightforward to show that the dot product is given by

$$\mathbf{R} \cdot \mathbf{S} = R_xS_x + R_yS_y + R_zS_z. \qquad \text{(A-4)}$$

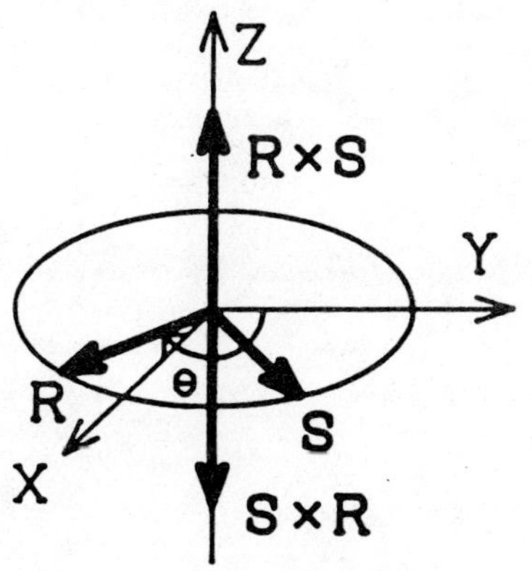

Figure A.1 The vector cross product of two vectors. Note that the vectors **RxS** and **SxR** are antiparallel; they are both perpendicular to the plane defined by **R** and **S**.

As the name implies, the scalar product, **R·S**, is a scalar. That is, it has a magnitude but no directional property. The vector product, or cross product, of two vectors oriented at an angle of θ to one another is a vector, thus, **RxS** = **T** (see Figure A.1). The magnitude of the cross product is given by

$$|\mathbf{R}\mathrm{x}\mathbf{S}| = |\mathbf{R}|\ |\mathbf{S}|\ \sin\theta\ . \qquad \text{(A-5)}$$

The direction of the cross product, **T**, is perpendicular to the plane defined by R and S. In this book we use the right hand convention which means that **RxS** points upward and **SxR** points downward (see Figure A.1). In any event, one may write

$$\mathbf{R}\mathrm{x}\mathbf{S} = -\ \mathbf{S}\mathrm{x}\mathbf{R}\ . \qquad \text{(A-6)}$$

The customary way to express a cross product in terms of the components of the components of **R** and **S** is to use the determinental relation

$$\mathbf{R} \times \mathbf{S} = \begin{vmatrix} \mathbf{i} & \mathbf{j} & \mathbf{k} \\ R_x & R_y & R_z \\ S_x & S_y & S_z \end{vmatrix}\ . \qquad \text{(A-7)}$$

In terms of the individual components we have

$$\mathbf{R} \times \mathbf{S} = \mathbf{i}(R_y S_z) + \mathbf{j}(R_z S_x) + \mathbf{k}(R_x S_y), \qquad \text{(A-8)}$$

or,

$$[\mathbf{R} \times \mathbf{S}]_x = R_y S_z$$
$$[\mathbf{R} \times \mathbf{S}]_y = R_z S_x \qquad \text{(A-9)}$$
$$[\mathbf{R} \times \mathbf{S}]_z = R_x S_y$$

REFERENCES

1) B.G. Lasarew and L.W. Schubnikow, Phys. Z. Sowjet 11, 445 (1937).

2) C.J. Gorter, Physica 3, 995, (1936).

3) E.M. Purcell, H.C. Torrey and R.V. Pound, Phys. Rev. 69, 37 (1946)

4) F. Bloch, W.W. Hanson, and M. Packard, Phys. Rev. 70, 474 (1946).

5) E.L. Hahn, Phys. Rev. 80, 580 (1950).

6) H.C. Torrey, Phys. Rev. 76, 1059 (1949).

7) R.R. Ernst and W.A. Anderson, Rev. Sci. Instr. 37, 93 (1966).

8) T. Keller, Am. Chem. Soc. Fifth Western Regional Meeting, Anaheim, CA., October 7, 1969.

9) T.C. Farrar and E.D. Becker, Am. Chem. Soc. Fifth Western Regional Meeting, Anaheim, CA., October 7, 1969.

10) A. Abragam, "The Principles of Nuclear Magnetism," Oxford Univ. Press, London and New York, (1961).

11) See, for example, ref. 10, pages 65,66.

12) C.P. Slichter, "Principles of Magnetic Resonance," Harper, New York, 1963.

13) P.C. Lauterbur, J. Amer. Chem. Soc. 83, 1838 (1961).

14) R.B. Blackman and J.W. Tukey, "The Measurement of Power Spectra," Dover, New York, (1958); J.W. Cooley and FJ.W. Tukey, Math. Comput. 19, 297 (1965).

15) R. Bracewell, "The Fourier Transform and Its Applications," pp. 6-7. McGraw-Hill, New York, 1965.

16) E.O. Stejskal and J.E. Tanner, J. Chem. Phys. 42, 288 (1965).

17) H.Y. Carr and E.M. Purcell, Phys. Rev. 94, 630 (1954).

18) R.L. Vold, R.R. Vold, and H.E. Simon, J. Mag. Res. 11, 283 (1973).

19) J.D. Ellett, Jr., M.G. Gibby, U. Haeberlen, L.M. Huber, M. Mehring, A. Pines, and J.S. Waugh, Adv. Mag. Res. 5 (1971).

20) A.G. Redfield, and R.K. Gupta, Adv. Mag. Res. 5, (1971); A.G. Refield and S.D. Kunz, J. Mag. Res. 19, 250 (1975).

21) R.W. Vaughan, D.D. Ellerman, L.M. Stacy, W.K. Rhim, and J.W. Lee, Rev. Sci. Instr. 43, 1356 (1972).

22) E.O. Stejskal and J. Schaefer, J. Mag. Res. 18, 560 (1975).

23) F.A.L. Anet, Topics C-13 NMR Spectros. 1 (1974).

24) G.C. Levy, I.R. Peat, R. Rosanske, and S. Parks, J. Mag. Res. 18, 205 (1975).

25) B.L. Tomlinson and H.D.W. Hill, J. Chem. Phys. 59, 1775 (1973).

26) R. Freeman and H.D.W. Hill, J. Mag. Res. 4, 336 (1971).

27) K.G.R. Packler and P.L. Wessels, J. Mag. Res. 12 (1973).

28) S. Sorenson, R.S. Hansen and H.J. Jacobsen, J. Mag. Res. 14, 243 (1974).

29) H.J. Jacobsen, S.A. Linde and S. Sorensen, J. Mag. Res. 15, 385

30) G.A. Morris, G.L. Nayler, A.J. Shaka, J. Keeler and R. Freeman, J. Magn. Reson. 58, 155 (1984); A.J. Shaka, J. Keeler, R. Freeman, G.A. Morris and G.L. Nayler, J. Magn. Reson. 58, 161 (1984).

31) J. S. Waugh, J. Magn. Reson. 49, 517 (1982). See also J.W.M. Jacobs, J.W.M. van Os, and W.S. Veeman, ibid., 51, 56 (1983).

32) R. Freeman, S.P. Kempsell and M.H. Levitt, J. Magn. Reson. 35, 447 (1979); M.H. Levitt and R. Freeman, ibid., 43, 502 (1982).

33) G.A. Morris and R. Freeman, J. Am. Chem. Soc. 101, 760 (1979).

34) D.P. Burum and R.R. Ernst, J. Magn. Reson. 39, 163 (1980).

35) A. Bax, R. Freeman, and S.P. Kempsel, J. Am. Chem. Soc. 102, 4849 (1980).

36) E.R. Andrew, "Nuclear Magnetic Resonance", Cambridge University Press, Cambridge, (1955).

37) N. Bloembergen, E.M. Purcell and R.V. Pound, Phys. Rev. 73, 679 (1948).

38) R.C. Tolman, "The Principles of Statistical Mechanics", Oxford University Press, London, (1962).

39) T.C. Farrar and R.A. Quintero, Chem. Phys. Letts. 122, 41 (1985); T.C. Farrar, B.R. Adams, G.C. Grey, R.A. Quintero-Arcaya, and Q. Zho, J. Am. Chem. Soc., 109, 0000 (1987).

40) T. Tsang and T.C. Farrar, J. Chem. Phys. 50, 3498 (1969).

41) W.B. Moniz and H.S. Gutowsky, J.Chem. Phys. 38, 1155 (1963).

42) T.C. Farrar, S.J. Druck, R.R. Shoup and E.D. Becker, J. Am. Chem. Soc. 94, 699 (1972).

43) J. Winter, C.R. Acad. Sci. Paris 249, 1346 (1959).

44) R.R. Vold and T. C. Farrar, unpublished work.

45) R.R. Shoup and D.L. VanderHart, J. Am. Chem. Soc. 93, 2053 (1971).

46) A. Allerhand, Rev. Sci. Instr. 41, 269 (1970).

47) A. Allerhand, and H.S. Gutowsky, J. Chem. Phys. 41, 2115 (1964); ibid. 42, 1587 (1965).

48) P.S. Hubbard, Phys. Rev. 131, 1155 (1963).

49) R.E.D. McClung, J. Chem. Phys. 51, 3842 (1969).

50a) A.A. Maryott, T.C. Farrar, and M.S. Malmberg, J. Chem. Phys. 54, 64 (1971). (b) T.C. Farrar and G.R. Quinting, J. Phys Chem. 90, 2834 (1986)

51) C. Deverell, Mol. Phys. 18, 319 (1970).

52) J.W. Cooper, "The Minicomputer in the Laboratory", John Wiley and Sons, New York (1977).

53) H. Nyquist, Trans. A.I.E.E., 617-644, April 1928.

53a) J.G. Ables, Proc. Symp. on Collection and Analysis of Astrophysical Data, 15, 383 (1972).

53b) C. Chatfield, "The Analysis of Time Series", Chapman and Hall, London (1984).

53c) H. Barkhuijsen, R. deBeer, W.M.M.J. Bovee, and D. vanOrmondt, J. Magn. Reson. 61, 465 (1985).

53d) H. Barkhuijsen, R. deBeer, and D. vanOrmondt, J. Magn. Reson. 64, 343 (1985).

54) E.D. Becker, J.A. Ferretti, and T.C. Farrar, J. Am. Chem. Soc. 91, 7784 (1969); see also R.L. Streever and H.Y. Carr, Phys. Rev. 121, 20 (1961).

55) J.S. Waugh, J. Mol Spectrosc. 35, 298 (1970).

56) E.B. Baker, J. Chem. Phys. 37, 911 (1963).

57) A.W. Overhauser, Phys. Rev. 92, 411 (1953).

58) R.A. Bell and J.K. Saunders, Can. J. Chem. 46, 3421 (1968); ibid. 48, 1114 (1970).

59) J.H. Noggle and R.E. Schirmer "The Nuclear Overhauser Effect", Academic Press, New York and London (1971).

60) K.G.R. Packer and P.L. Wessels, J. Magn. Reson. 12, 337 (1973).

61) G.A. Morris and R. Freeman, J. Am. Chem. Soc. 101, 760 (1979).

62) G.A. Morris, J. Am. Chem. Soc. 102, 428 (1980).

63) D.M. Doddrell and D.T. Pegg, J. Am. Chem. Soc. 102, 6388 (1980).

64) P.H. Bolton, J. Magn. Reson. 41, 287 (1980).

65) A.A. Maudsley and R.R. Ernst, Chem. Phys. Lett. 50, 368 (1977).

66) D.P. Burum and R.R. Ernst, J. Magn. Reson. 39, 163 (1980).

67) D.M. Doddrell, D.T. Pegg and M.R. Bendall, J. Magn. Reson. 48, 323 (1982).

68) G. Bodenhausen, Prog. NMR Spectros. 14, 137 (1981).

69) J. Jeener and G. Alewaeters, unpublished work; J. Jeener, Ampere Int'l. Summer School, Yugoslavia, (1971), unpublished; G. Alewaeters, Doctoral Thesis, Free University of Brussels (1976).

70) W.P. Aue, E. Bartholdi and R.R. Ernst, J. Chem. Phys. 64, 2229 (1976).

71) L. Mueller, A. Kumar, and R.R. Ernst, J. Chem. Phys. 63, 5490 (1975).

72) G. Bodenhausen, R. Freeman, R. Niedermeyer, D.L. Turner, J. Magn. Reson. 24, 291 (1976)

73) An excellent review of the entire subject of 2D-FT-NMR is given in the book by Dr. Ad Bax "Two-Dimensional Nuclear Magnetic Resonance in Liquids", D. Reiden, Dordrecht (1982).

74) A. Bax, R. Freeman, J. Magn. Reson. 42, 164 (1981).

75) K. Nagayama, A. Kumar, K. Wuethrich, and R.R. Ernst, J. Magn. Reson. 40, 321 (1980).

76) D.F. Gaines, G. Edvensen, T. Hill and B.R. Adams, Inorg. Chem. 00, 0000, 1987.

SUBJECT INDEX

D

E

F

N

P

Q

R

Z